W9-AUP-206

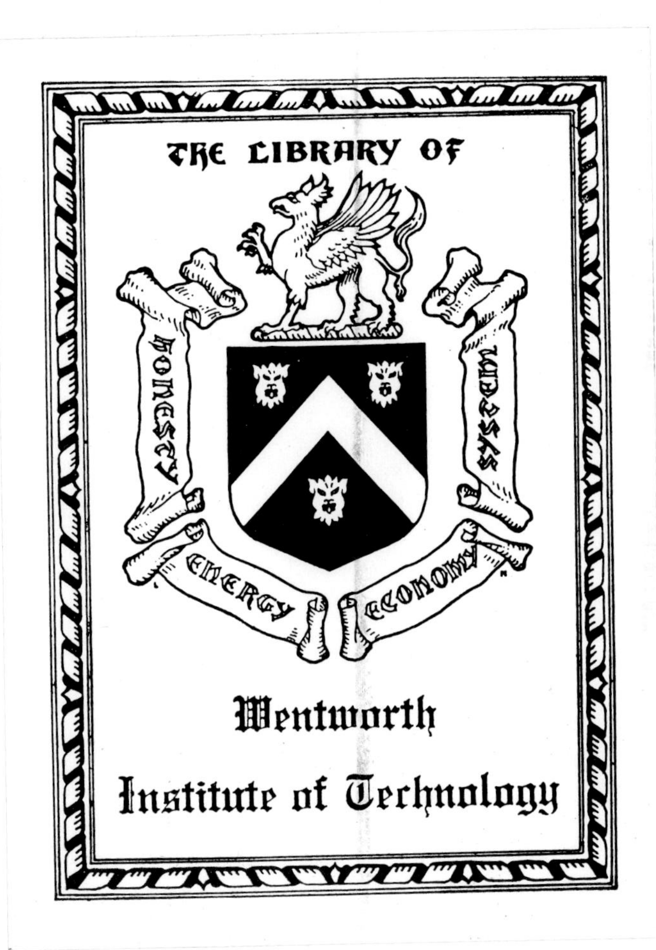
THE LIBRARY OF
HONESTY
SYSTEM
ENERGY
ECONOMY
Wentworth
Institute of Technology

The quantum universe

Tony Hey, READER,
PHYSICS DEPARTMENT, SOUTHAMPTON UNIVERSITY

Patrick Walters, LECTURER,
ADULT AND CONTINUING EDUCATION DEPARTMENT,
UNIVERSITY COLLEGE OF SWANSEA

CAMBRIDGE UNIVERSITY PRESS
Cambridge
London New York New Rochelle
Melbourne Sydney

Published by the Press Syndicate of the University of Cambridge
The Pitt Building, Trumpington Street, Cambridge CB2 1RP
32 East 57th Street, New York, NY 10022, USA
10 Stamford Road, Oakleigh, Melbourne 3166, Australia

First published 1987

Printed in Great Britain by Scotprint Ltd

British Library cataloguing in publication data

Hey, Anthony J. G.
The quantum universe.
1. Quantum theory
I. Title II. Walters, Patrick
530.1′2 QC174.12

Library of Congress cataloguing in publication data

Hey, Anthony J. G.
The quantum universe.

Bibliography
Includes indexes.

1. Quantum theory. I. Walters, Patrick,
1949– . II. Title.
QC174.12.H48 1987 530.1′2 86–6830

ISBN 0 521 26744 7 hard covers
ISBN 0 521 31845 9 paperback

SE

CONTENTS

PREFACE

The goal of this book is to present the essential ideas of quantum physics as simply as possible and demonstrate how quantum physics affects us all in our everyday life. We therefore do not focus our discussion on philosophical debates about the meaning of quantum mechanics. Rather, we describe how the seemingly bizarre and 'academic' notions of quantum mechanics have been successfully applied to an enormously diverse range of fields. Some idea of the scope of these applications can be gained from the contents list of this book.

There is, however, one formidable problem to be faced before such a 'popular' approach to quantum mechanics can become a reality. The problem is this. The natural language in which to express the ideas of physics is mathematics. For this reason, understanding physics, and quantum mechanics in particular, presents those without the necessary mathematical background with a huge obstacle to be overcome. Nonetheless, the impact of quantum mechanics on our lives is so enormous that we firmly believe that almost any attempt to bridge the gulf between the literate and the numerate is worthwhile. To quote from G. K. Chesterton: 'If a thing is worth doing, it's worth doing badly!' It is therefore at the risk of accumulating considerable scorn from some of our academic colleagues, that we attempt in this book to explain the whole of quantum physics without recourse to mathematics or equations!

Our approach is based on the belief that once some idea of the fundamental peculiarity of quantum motion – crudely speaking, that 'particles' can behave like 'waves' – has been gained, then much that was formerly incomprehensible about quantum mechanics becomes 'obvious'. In this way, the simple example of waves on a string leads directly to 'energy quantization', and thence to much of atomic and nuclear physics. Similarly, the strange phenomenon of quantum tunnelling, responsible for nuclear fission, hydrogen burning in the stars, and much else besides, is directly related to similar behaviour observed for water waves. The surprises come only from the basic wave–particle duality of quantum mechanics.

To make clear our goal that this book should be as close to a 'coffee-table quantum mechanics' book as we can make it, we have toyed with many possible titles. At various times we have referred to the book as *Quantum Mechanics for Bank Managers* (financial expediency), *Quantum Mechanics for Housewives* (frankly sexist), *Quantum Mechanics for Builders* (evidently practical), and so on. Perhaps most relevant would be a title such as *Quantum Mechanics for Politicians*. It is a truly appalling comment on our education system that those who guide our economic destiny often have little or no conception of the impact that quantum mechanics and 'basic physics' research has had, and continues to have, on our modern way of life. One example will suffice to make the point. In conversation, the research director of the 'high-tech' division of one of the UK's leading companies dismissed quantum mechanics as being irrelevant to the physics observed at ordinary 'room' temperatures. He believed that some sort of classical explanation would always be adequate. In fact, from the standpoint of classical physics even the floor that he was standing on, let alone his industry, would not exist! This type of statement, however, seems to typify the feeling prevalent among industrialists and government today that investment in 'basic science' is less beneficial than investment in 'applied' research. One of the physicists mentioned in this book, Henrick Casimir, is exceptional in that he not only made contributions to our understanding of quantum mechanics but also became director of Philips Research Laboratories in Holland, and therefore had experience of both academic and industrial research. Casimir's refutation of this viewpoint is memorable:

I have heard statements that the role of academic research in innovation is slight. It is about the most blatant piece of nonsense that it has been my fortune to stumble upon.

Certainly, one may speculate idly whether transistors might have been discovered by people who had not been trained in and had not contributed to wave mechanics or the theory of electrons in solids. It so happened that inventors of transistors were versed in and contributed to the theory of solids.

One might ask whether basic circuits in computers might have been found by people who wanted to build computers. As it happens, they were discovered in the thirties by physicists dealing with the counting of nuclear particles because they were interested in nuclear physics. One might ask whether there would be nuclear power because people wanted new power sources or whether the urge to have new power would have led to the discovery of the nucleus. Perhaps – only it didn't happen that way, and there were the Curies and Rutherford and Fermi and a few others.

One might ask whether an electronic industry could exist without the previous discovery of electrons by people like Thomson and H. A. Lorentz. Again it didn't happen that way.

One might ask even whether induction coils in motor cars might have been made by enterprises which wanted to make motor transport and whether they would have stumbled on the laws of induction. But the laws of induction had been found by Faraday many decades before that.

Or whether, in an urge to provide better communication, one might have found electromagnetic waves. They weren't found that way. They were found by Hertz who emphasized the beauty of physics and who based his work on the theoretical considerations of Maxwell. I think there is hardly any example of twentieth century innovation which is not indebted in this way to basic scientific thought.

This point of view was also shared by men like Faraday and J. J. Thomson. Faraday, when asked by the famous prime minister Gladstone as to the practical use of the discovery of electricity replied: 'One day, Sir, you may tax it'. Similarly, Thomson, who discovered the electron, remarked that while research into applied science leads to improvement and development of older methods, research in pure science can result in entirely new and more powerful methods. He concluded that 'research in applied science leads to reform, research in pure science leads to revolutions, and revolutions, whether political or industrial, are exceedingly profitable things if you are on the winning side'.

Our book can therefore be seen as a defence of basic science – but that was not our real reason for writing it. We wrote it because we ourselves find the quantum universe a source of endless fascination, and we would like to bring a sense of this excitement to as wide an audience as possible. We hope that our book will stimulate young people to find out more, and discover the true power of quantum mechanics obtained by adjoining mathematics to the qualitative descriptions given in this book. We also hope that it will appeal to older readers who wish to know something about the way in which the quantum world appears to work. We believe it desirable that more people should understand what physics can and cannot do, and appreciate how physics has made possible the new 'high technologies' that are changing our lives.

This book has grown out of lectures given in Southampton and Swansea Universities and we wish to acknowledge all the valuable suggestions made by many of our loyal students. We also thank many of our friends and colleagues for their interest and help. In particular, Tony Hey wishes to thank Ian Aitchison and Malcolm Coe for helpful comments, Tessa Coe and Charlie Askew for help with computer graphics, and, most of all, Garry McEwen, for his tireless assistance throughout this project without which this book would be much more obscure. Patrick Walters also thanks Colin Grey Morgan, Steve Hibbs, Ann Jenkins and Howard Miles for their help in making this book what it is. Needless to say, the remaining flaws are all ours. Last, but obviously not least, we wish to thank Jessie and Marie for their tolerance and forebearance throughout what was intended as a very short project but ended up as a very much longer one!

We also wish to record our thanks to all the staff of Cambridge University Press who have contributed in producing a book so close to our original conception. We are particularly indebted to Simon Capelin and Robin Rees for not only having faith in the project but also seeing us through the difficult times with good humour and patience. Irene Pizzie and Jeanette Hurworth also made invaluable contributions to our preparation of the text and photographs for publication. To them all, many thanks.

One final 'thank you' is in order. The reader may wonder why we have chosen to introduce each chapter with a quotation from Richard Feynman. The informal flavour of the quotations gives some indication of Feynman's non-pompous, plain-speaking style of communication. Moreover, Feynman is without equal both as a research physicist and as an educator. He is a passionate believer in the importance of understanding as opposed to ritual learning of formulae and definitions. For anyone not familiar with his career and personality, his recent collection of anecdotes *Surely You're Joking, Mr. Feynman!* makes absorbing reading. Feynman's style of teaching is truly inimitable, combining as it does, great understanding of the physics with great originality of presentation. In this book, we try to follow Feynman's example, and hope that some of the freshness of his style of presentation comes through in our text. In a very real sense, therefore, this book is dedicated to Richard Feynman, for all that he has given to physics. We are not at all sure that he will like the results of our labours: one thing we are sure of is that he would have done it differently!

PROLOGUE

Poets say science takes away from the beauty of the stars – mere globs of gas atoms. Nothing is 'mere'. I too can see the stars on a desert night, and feel them. But do I see less or more? The vastness of the heavens stretches my imagination – stuck on this carousel, my little eye can catch one-million-year-old light . . . Or see them [the stars] with the greater eye of Palomar, rushing all apart from some common starting point when they were perhaps all together. What is the pattern, or the meaning, or the why? It does not do harm to the mystery to know a little about it. For far more marvellous is the truth than any artists of the past imagined! Why do the poets of the present not speak of it?

Richard Feynman

ROUTE MAP

The diagram below shows the major interconnections of the various chapters and topics covered in the book.

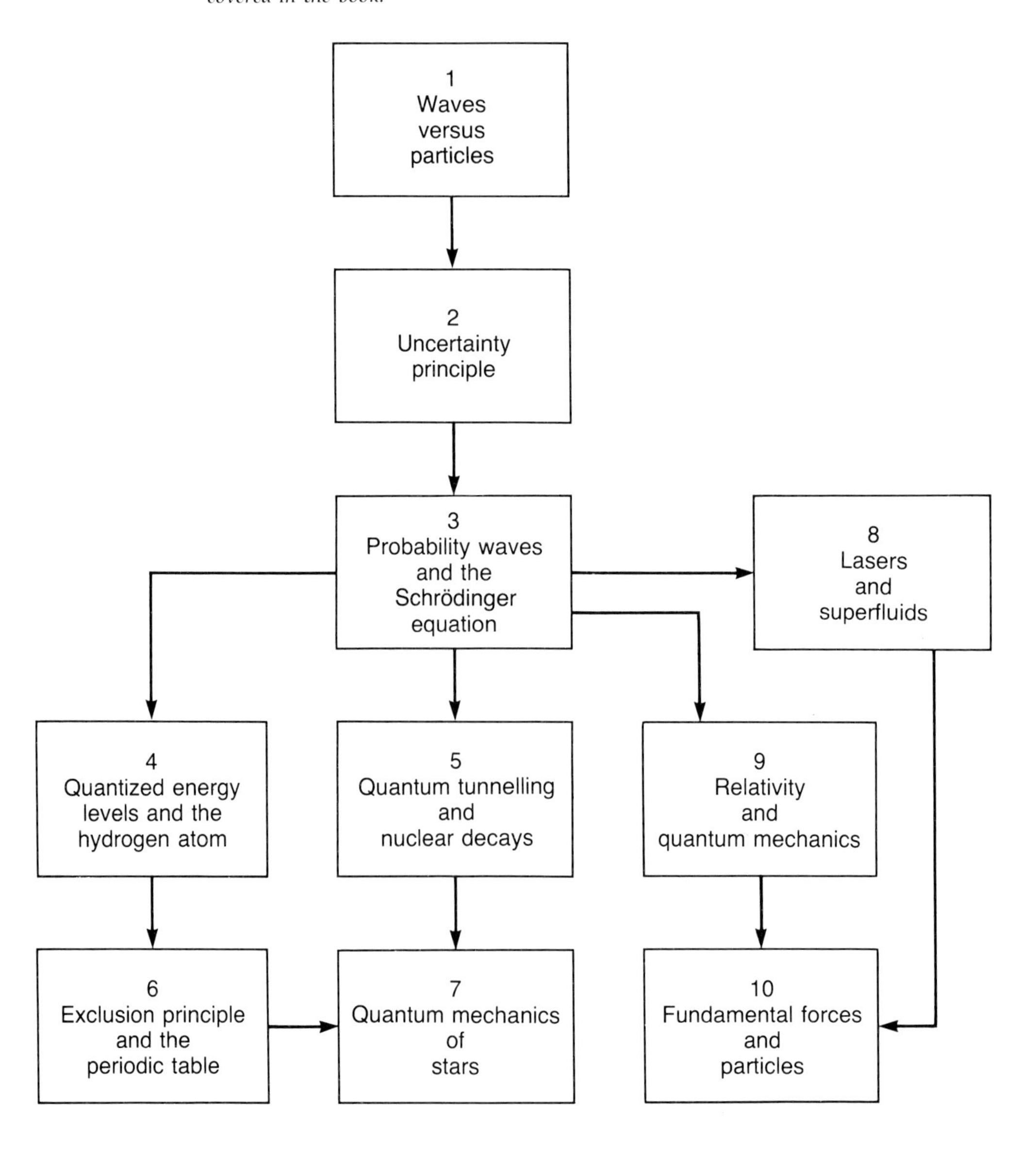

1

Waves versus particles

. . . I think I can safely say that nobody understands quantum mechanics.
Richard Feynman

Science and experiment

Science is a special kind of explanation of the things we see around us. It starts with a problem and curiosity. Something strikes the scientist as odd. It doesn't fit in with the usual explanation. Maybe harder thinking or more careful observation will resolve the problem. If it remains a puzzle, it stimulates the scientist's imagination. Perhaps a completely new way of looking at things is needed? Scientists are perpetually trying to find better explanations – better in the sense that any new explanation must not only explain the new puzzle, but also be consistent with all of the previous explanations that still work well. The hallmark of any scientific explanation or 'theory' is that it must be able to make successful predictions. In other words, any decent theory must be able to say what will happen in any given set of circumstances. Thus, any new theory will only become generally accepted by the scientific community if it is able, not only to explain the observations that scientists have already made, but also to foretell the results of new, as yet unperformed, experiments. This rigorous testing of new

Isaac Newton (1642–1727) published his book Optics *in 1704 that explained the rainbow and put forward the 'corpuscular' theory of light. In his book* Mathematical Principles of Natural Philosophy *Newton set down the principles of mechanics and gravity that guided science until the mid 19th century.*

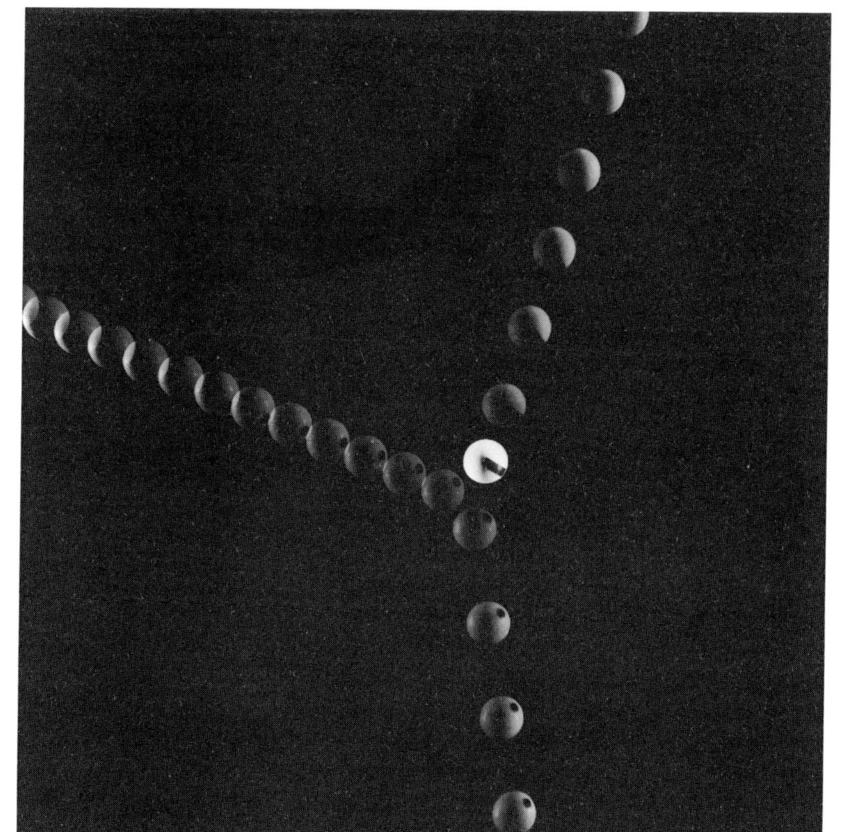

Fig. 1.1 A multi-flash photograph of a billiard ball collision. The motions of the balls can be calculated using Newton's laws but we have a good feel for what will happen from watching snooker on television or playing ourselves.

Fig. 1.2 Astronaut Bruce McCandless floats in space during the first untethered space walk on February 7th, 1984. The astronaut is essentially an independent space-craft in orbit near the shuttle. McCandless commented 'Well that may have been one small step for Neil [Armstrong] but it's a heck of a big leap for me!'

scientific ideas is the key feature that distinguishes science from other fields of intellectual endeavour – such as history or even economics – or from a pseudoscience such as astrology.

In the 17th century Isaac Newton and several other great scientists developed a wonderfully successful explanation of the way things move. This whole theoretical framework is called 'classical mechanics', and its scope encompasses the motion of everything from billiard balls to planets. Newton's explanation of motion in terms of forces, momentum and acceleration is encapsulated in his 'laws of motion'. These principles are incorporated into so many of our machines and toys that classical mechanics is familiar from our everyday experience. We all know what to expect in the collision of two billiard balls. Perhaps the most spectacular application of classical mechanics is in the exploration of space. Nowadays it surprises no one that the astronaut and the space shuttle float side by side and neither falls dramatically to Earth. A hundred years ago it was not so 'obvious', and in Jules Verne's famous story *A Trip Around The Moon* the passengers of the spacecraft were amazed to find the body of a dog that died on takeoff, and which they had jettisoned outside the craft, floating side by side with them all the way to the Moon. Today, you may not know how Newton's theory works in detail but you can see that it works. It is part of our daily experience.

All this brings us to the problem most of us have in coming to terms with 'quantum mechanics'. It is just this. At the very small distances involved in the study of atoms and molecules, things do *not* behave in a familiar way. Classical mechanics is inadequate and an entirely new explanation is needed. Quantum mechanics is that new explanation, and it is cunningly

Fig. 1.3 In the story by Jules Verne A Trip Around The Moon, *published in 1865, the dog 'Satellite' died on take-off and was jettisoned from the space-ship. Much to the surprise of the occupants the dog's body floated along with them all the way to the Moon!*

constructed so that it not only works in the quantum domain of very short length scales, but also so that, for larger distances, its predictions are identical with those of Newton. An atom is a typical quantum thing – it cannot be understood from the standpoint of classical physics. One popular visualization of an atom imagines electrons orbiting the nucleus of the atom much in the way planets orbit the Sun in the solar system. In fact, for negatively charged electrons in orbit round a positively charged nucleus, this simple model is unstable! According to classical physics the electrons would spiral into the centre and the atom would collapse. This nice and comforting model of the atom simply cannot account for even the existence of real atoms, let alone predict their expected behaviour. It is important to be aware at the outset that there is *no* simple picture that can accurately describe the behaviour of electrons in atoms. This is the first hurdle faced by the newcomer to the quantum domain: the inescapable and unpalatable fact that behaviour of quantum objects is totally unlike anything you have ever seen.

How can we convince you that quantum mechanics is both necessary and useful? Well, a physicist, just like a good detective, sifts through the evidence and remembers the old maxim of Sherlock Holmes that 'when you have excluded the impossible, whatever remains, however improbable, must be the truth'. Nonetheless, it was not without much reluctance that 20th century physicists became convinced that the whole magnificent edifice of classical physics was not 'almost right' for describing the behaviour of atoms, but had, instead, to be radically rebuilt. Nowhere was the confusion generated by this painful realization more evident than in their attempts to understand the nature of light.

Light and quantum mechanics

It was way back in the 17th century that Isaac Newton suggested that light should be regarded as a stream of particles, rather like bullets from a machine gun. Such was Newton's reputation that this view persisted, apart from some isolated pockets of opposition, until the 19th century. It was then that Thomas Young and others conclusively showed that the particle picture of light must be wrong. Instead, they favoured the idea that light was a kind of wave motion. One property of waves that is familiar to us is that of 'interference', to use the physicists' term for what happens when two waves collide. For example, in fig. 1.4 we show the 'interference' patterns produced by two sources of water waves on the surface of the water. Using his famous 'double-slit' apparatus to make two sources of light, Thomas Young had observed similar interference patterns using light.

Thomas Young (1773–1829) was an infant prodigy who could read at age two. During his youth he learnt to speak a dozen languages. He is best remembered for his work on vision and for establishing the wave theory of light. However, he was also the first to make progress on deciphering the hieroglyphic language of the ancient Egyptians.

Alas, physicists were not able to congratulate themselves for long. Experiments at the end of the 19th century revealed effects that were inexplicable by a wave theory of light. The most famous of such experiments concerns the so-called 'photo-electric' effect. Ultra-violet light shone onto a negatively charged metal caused it to lose its charge, while shining visible light on the metal had no effect. This puzzle was first explained by Albert Einstein in the same year that he invented the 'theory of relativity' for which he later became famous. His explanation of the photo-electric effect resurrected the particle view of light. The discharging of the metal was caused by electrons being knocked out of the metal by light energy concentrated into individual little 'bundles' of energy, which we now call 'photons'. According to Einstein's theory, ultra-violet photons have more energy than visible-light ones, and so no matter how much visible light you shine on the metal, none of the photons have enough energy to kick out an electron.

After several decades of confusion in physics, a way out of this mire was found in the 1920s with the emergence of quantum mechanics, pioneered

Fig. 1.4 Photograph of the interference pattern produced by two vibrating sources in water.

by physicists such as Heisenberg, Schrödinger and Dirac. This theory is able to provide a successful explanation of the paradoxical nature of light, atoms and much else besides. But there is a price to pay for this success. We must abandon all hope of being able to describe the motion of things at atomic scales in terms of everyday concepts like waves or particles. A 'photon' does not behave like anything anyone has ever seen. This does not, however, mean that quantum mechanics is full of vague ideas and lacks predictive power. On the contrary, quantum mechanics is the only theory capable of making definite and successful predictions for systems of atomic sizes or smaller in much the same way that classical mechanics makes predictions

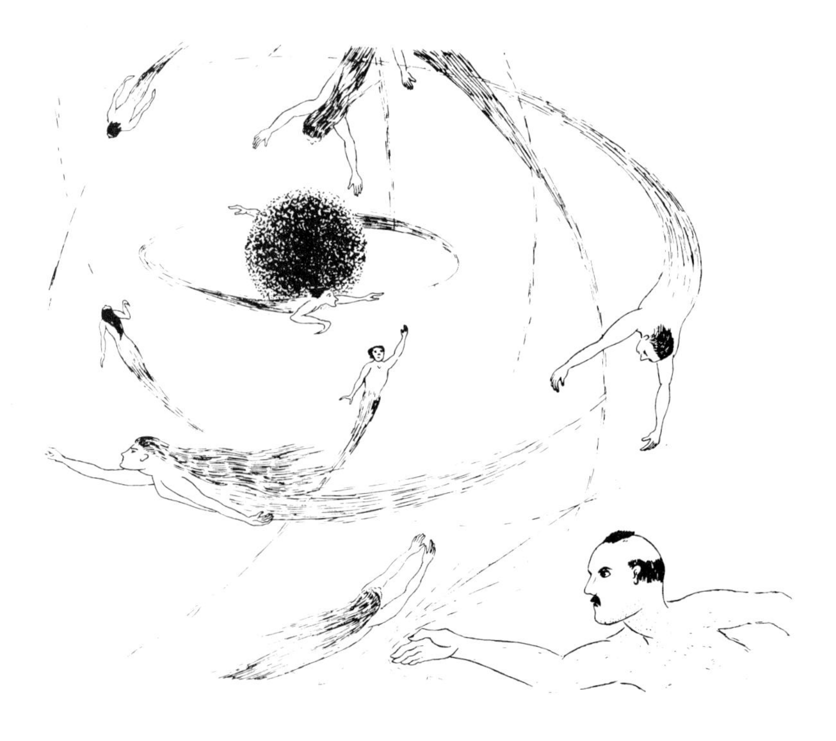

Fig. 1.5 George Gamow's rather whimsical view of the planetary model of the atom in Mr Tompkins Explores The Atom.

J. J. Thomson (1856–1940) measured the charge-to-mass ratio of the electron thus establishing it as a new elementary particle. He was awarded the Nobel Prize in 1906.

for the behaviour of billiard balls, rockets and planets. The difficulty with quantum things such as the photon is that, unlike billiard balls, their motion cannot be visualized in any accurate pictorial way. All we can do is summarize our lack of a picture by saying that a photon behaves in an essentially quantum mechanical way.

There is one sense in which Nature has been kind to us. Viewed from the perspective of classical physics, photons and electrons seem like very different kinds of objects. Remarkably, in the quantum domain both photons and electrons, and indeed all quantum objects, behave in the same strange quantum mechanical way. This is at least some compensation for our inability to picture quantum things! There is a curious little irony in the history of our attempts to understand the nature of electrons. In 1897 J. J. Thomson measured the charge-to-mass ratio of the electron and established the electron as a new elementary particle of Nature. Thirty years later, his son, G. P. Thomson, and also Davisson and Germer in the USA, performed a beautiful series of experiments which conclusively revealed that electrons also behave like waves. The historian Max Jammer wrote: 'One may feel inclined to say that Thomson, the father, was awarded the Nobel Prize for having shown that the electron is a particle, and Thomson, the son, for having shown that the electron is a wave'.

Our intention in this book is to impress even the most skeptical reader with the enormous range and diversity of the successful predictions of quantum mechanics. The apparently absurd ideas of de Broglie, Schrödinger and Heisenberg have now led to whole new technologies, the very existence of which depends on the discoveries of these pioneers of quantum mechanics. The modern electronics industry, with its silicon chip technology, is all based on the quantum theory of materials called semiconductors. Likewise, all the multitude of applications of lasers are possible only because of our understanding, at the fundamental quantum level, of a mechanism for radiation of light from atoms first identified by Einstein in 1916. Moreover, understanding how large numbers of quantum objects behave when packed tightly together leads to an understanding of all the different types of matter ranging from 'superconductors' to 'neutron stars'. In addition, although originally invented to solve fundamental problems concerned with the existence of atoms, quantum mechanics was found to apply with equal success to the tiny nucleus at the heart of the atom, and this has led to an understanding of radioactivity and nuclear reactions. As everyone knows, this has been a mixed blessing. Not only do we now know what makes the stars shine, but we also know how to destroy all of civilization with the awesome power of nuclear weapons.

Max Born (1882–1970) was awarded the 1954 Nobel Prize as a very belated recognition of his work on the probability interpretation of quantum mechanics. Born left Germany when Hitler came to power and was Professor of Natural Philosophy in Edinburgh from 1936 until his retirement in 1953.

But before we can explain how quantum mechanics made all these things possible, we must first attempt to describe the strange quantum mechanical behaviour of objects at atomic distance scales. This task is clearly difficult given the absence of any accurate analogy for the mathematical description of quantum behaviour. However, we can make progress if we use a mixture of analogy and contrast. Young's original 'double-slit' experiment used a screen with two slits in it to make two sources of light which could interfere and produce his famous 'interference fringes' – alternating light and dark lines. We shall describe the results of similar 'double-slit' experiments carried out using bullets, water waves and electrons. By comparing and contrasting the results obtained with the three different materials we shall be able to give you some idea of the essential features of quantum mechanical behaviour. Quantum mechanics textbooks contain detailed discussion of many types of experiments, but this double-slit experiment is sufficient to reveal all the mystery of quantum mechanics. All of the problems and paradoxes of quantum physics can be demonstrated in this single experiment.

A word of warning before we begin. To avoid running into a frustrating psychological cul-de-sac, try to be content with mere acceptance of the observed experimental facts. Try not to ask the question 'but how can it be like that?' As Richard Feynman says 'nobody understands quantum mechanics'. All we can give you is an account of the way Nature appears to work. Nobody knows more than that.

The double-slit experiment

This section may be rather hard going first time through. If so, just glance at the pictures and pass on quickly to the next chapter!

WITH BULLETS

Source: a wobbly machine gun that, as it fires, spreads the bullets out into a cone, all with the same speed but random directions.
Screen: armour plate with two parallel slits in it.
Detector: small boxes of sand to collect the bullets.
Results: the gun fires at a fixed rate and we can count the number of bullets that arrive in any given box in a given period of time. The bullets that go through the slits can either go straight through or else bounce off one of the edges, but must always end up in one of the boxes. The bullets we are using are made of a tough enough metal so that they never break up – we can never have half a bullet in a box. Moreover, no two bullets ever arrive at the same time – we have only one gun, and each bullet is a single identifiable 'lump'.

If we let the experiment run for an hour and then count the bullets in each of the boxes, we can see how the 'probability of arrival' of a bullet varies with the position of the detector box. The total number of bullets arriving at any given position is clearly the sum of the number of bullets going through slit 1 plus the number going through slit 2. How this 'probability of arrival' varies with position of the sand boxes is shown in fig. 1.7. We shall label this result,

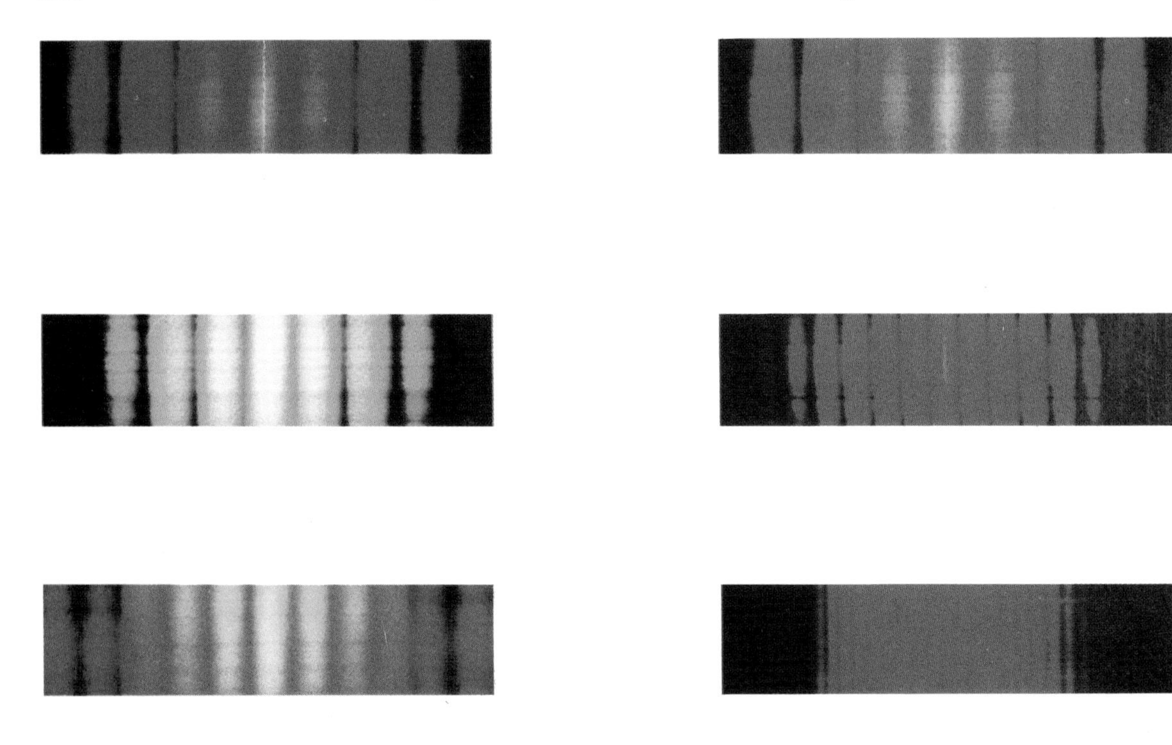

Fig. 1.6 Double-slit interference patterns for light, usually taken as demonstrating that light is a wave motion. In the left hand pictures, as the wavelength of the light is decreased and the colour changes from red to blue, the interference fringes become closer together. On the right, for red light, the decrease in the fringe separation is caused by increasing the separation of the slits.

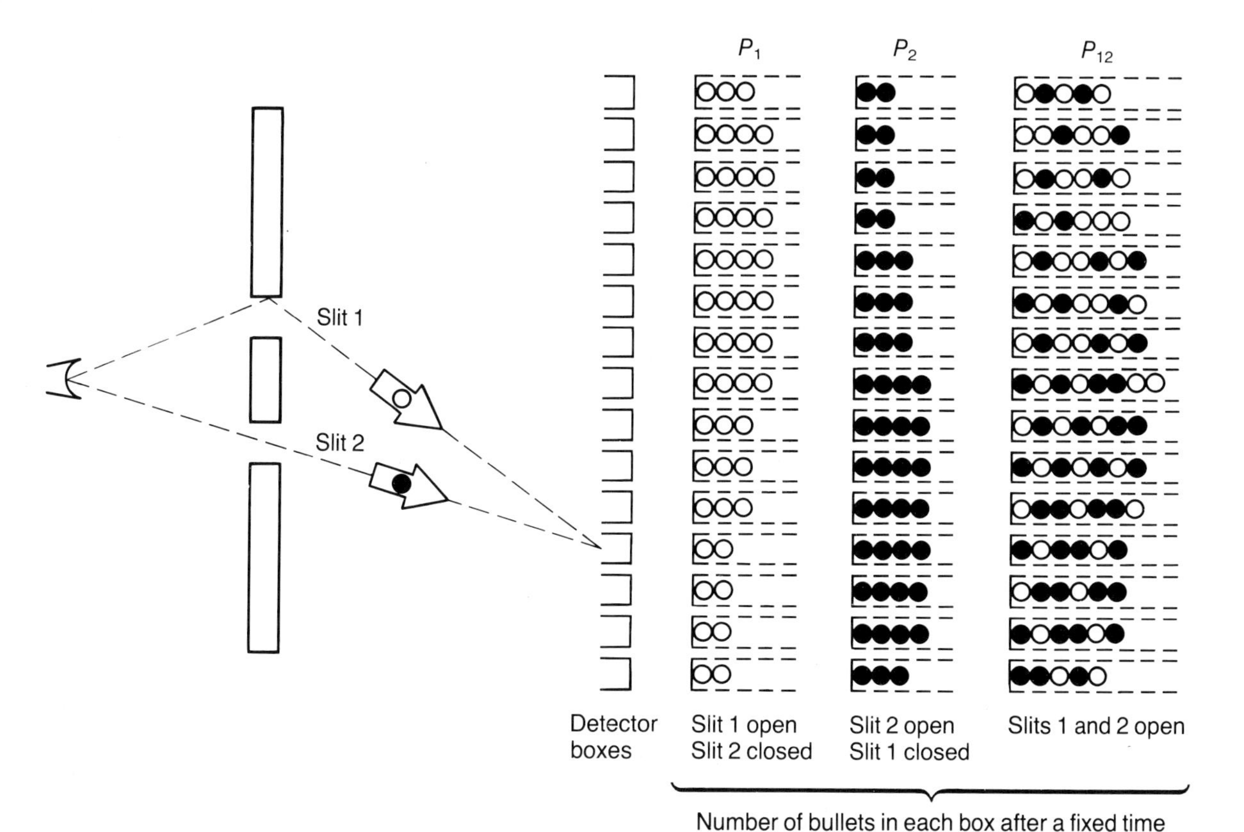

Fig. 1.7 A diagram of a double-slit experiment with bullets. The experimental set-up is shown on the left of the figure and the results of three different experiments indicated on the right. We have shown bullets that pass through slit 1 as open circles and bullets through slit 2 as black circles. The column labelled P_1 *shows the distribution of bullets arriving at the detector boxes when slit 2 is closed and only slit 1 is open. Column* P_2 *shows a similar distribution obtained with slit 1 closed and slit 2 open. As can be seen, the maximum number of bullets appears in the boxes directly in line with the slit that is left open. The result obtained with both slits open is shown in the column labelled* P_{12}*. It is now a matter of chance through which slit a bullet will come and this is shown by the scrambled mixture of black and white bullets collected in each box. The important point to notice is that the total obtained in each box when both slits are open is just the sum of the numbers obtained when only one or other of the slits is open. This is obvious in the case of bullets since we know that bullets must pass through one of the slits to reach the detector boxes.*

the probability of arrival of bullets when both slits are open, P_{12}. We also show in fig. 1.7 the results obtained with slit 2 closed, which we call P_1, and those obtained with slit 1 closed, which we call P_2. Looking at the figures, it is evident that the curve labelled P_{12} is obtained by adding curves P_1 and P_2. We can write this mathematically as the equation

$$P_{12} = P_1 + P_2$$

For reasons that will become apparent in a moment, we call this result the case of *no interference*.

WITH WATER WAVES

Source: a stone dropped into a large pool of water.
Screen: a jetty with two gaps in it.
Detector: a line of small floating buoys whose jiggling up and down gives a

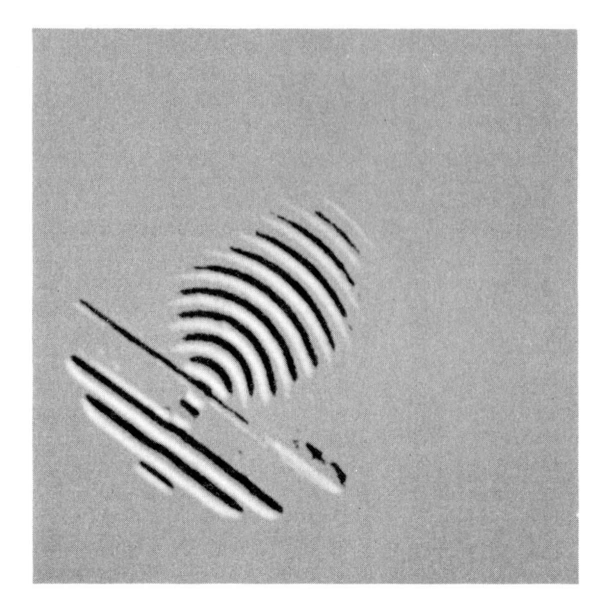

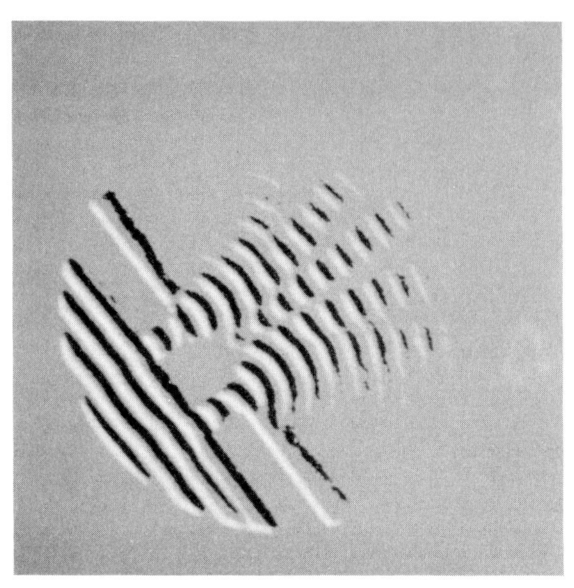

Fig. 1.8 Wave patterns with water waves. The left hand picture shows a wave spreading out from a single slit while the picture on the right shows the interference obtained with two slits.

measure of the amount of energy of the wave at that position.

Results: Ripples spread out from the source and reach the jetty. On the far side of the jetty ripples spread out from each of the gaps. At the detector, the resulting disturbance of the water is given by the sum of the disturbances of the ripples coming from both gaps. As we look along the line of buoys, there will be some places where the crest of a wave from slit 1 coincides with the arrival of a crest from slit 2, resulting in a very large up and down motion for the buoy. At other places, however, a crest from one slit will coincide with a trough from the other and there will be no movement of the buoy at that position. At other places the motion of the buoys will be somewhere between these two extremes. For water waves, it should be fairly plausible that the energy of a wave at any given position will be related to how big the waves are at that point. In fact, it can be shown that the energy of a wave depends on the square of the maximum height of the wave. Let us call the amount of energy arriving per second the 'intensity' and label this by the symbol I. If we label the maximum height of the wave by h, we can write the relation between I and h as the following equation:

$$I = h^2$$

intensity = height squared

In contrast to our experiment with bullets, we see that the energy of the waves does not arrive at the detector in definite sized lumps, like the bullets which only arrived at one particular position at one particular time. Rather, since the height of the resulting wave at the detector varies smoothly from zero up to some maximum value, we see that the energy of the original wave is spread out along the detector. The curve showing how the intensity varies with position along the detector is shown in fig. 1.9. Since this is the intensity obtained with both slits open, we shall call this curve I_{12}. This intensity pattern has a very simple mathematical explanation. The total disturbance of the water at any position along the detector is given by the sum of the disturbances caused by the waves from slit 1 and slit 2. If we label the height of the wave from slit 1 by h_1, the height from slit 2 by h_2, and the total height obtained when both slits are open as h_{12}, we can write this result as the equation

$$h_{12} = h_1 + h_2$$

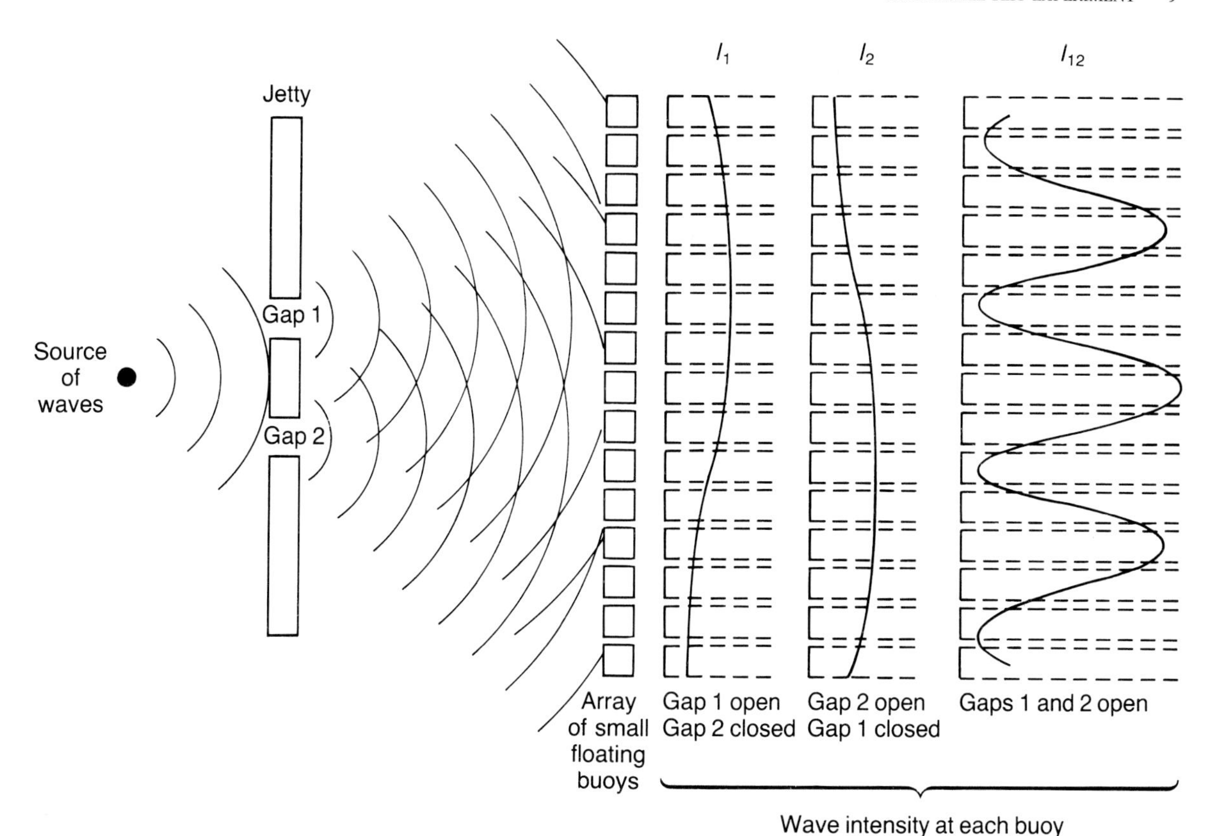

Fig. 1.9 A diagram of a double-slit experiment with water waves. The detectors are a line of small floating buoys whose jiggling up and down provides a measure of the wave energy. The wave crests spreading out from each slit are shown in the figure and can be compared with fig. 1.8. The column labelled I_1 *shows the smoothly varying wave intensity obtained when only gap 1 is open. Notice that this is very similar to the pattern* P_1 *obtained with bullets in fig. 1.7 with only slit 1 open. Again it is largest at the detector directly in line with gap 1 and the source. The second column shows that a similar pattern,* I_2*, is obtained when gap 1 is closed and gap 2 is open. The final column,* I_{12}*, shows the wave intensity pattern obtained with both slits open. It is dramatically different from the pattern obtained for bullets with both slits open. It is not equal to the sum of the patterns* I_1 *and* I_2 *obtained with one of the gaps closed. This rapidly varying intensity curve is called an interference pattern.*

Remember that each of these heights can be positive or negative depending on whether the corresponding wave disturbance raises or lowers the water level. The resulting intensity is just the square of this height or 'wave amplitude'

$$I_{12} = h_{12}^2$$

so that

$$I_{12} = (h_1 + h_2)^2$$

We could now repeat the experiment with one of the gaps closed. In this case we find the results shown in fig. 1.9. We label the corresponding intensity pattern I_1, since it is the intensity obtained with slit 1 open and slit 2 closed. The curve I_1 is just given by the square of the disturbance caused by the wave from slit 1

$$I_1 = h_1^2$$

Fig. 1.10 An experimental observation of the two-slit interference pattern obtained with electrons.

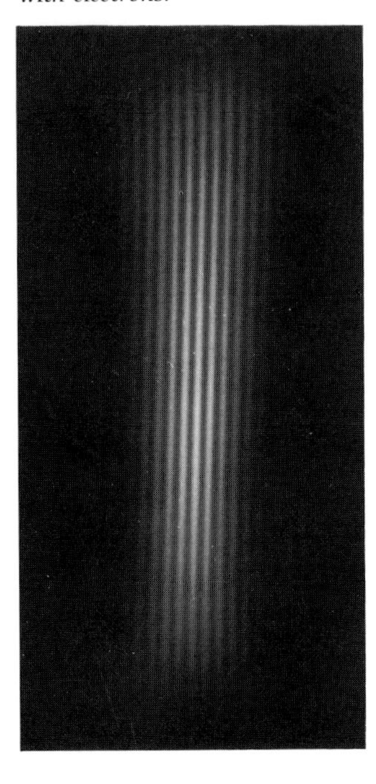

Similarly, the curve I_2 is the result obtained with slit 2 open and slit 1 closed, and, in the same way as before, we have the result

$$I_2 = h_2^2$$

It is clear that these two curves are much less wiggly than the pattern I_{12}. Furthermore, the pattern I_{12}, for both slits open, cannot be obtained just by adding up the two intensity patterns, I_1 and I_2, obtained with one of the slits closed. Mathematically, we can see this from our equations as follows:

$$I_{12} = (h_1 + h_2)^2$$
$$= (h_1 + h_2) \times (h_1 + h_2)$$

This may be expanded to read

$$I_{12} = h_1^2 + 2h_1h_2 + h_2^2$$

which is clearly not equal to the sum of I_1 and I_2

$$I_1 + I_2 = h_1{}^2 + h_2{}^2$$

For wave motion, we say there is *interference*. Unlike the case with bullets, you do not obtain the pattern for 'both slits open' by adding the patterns for 'one slit closed'. It was the observation of such interference patterns for light that convinced Thomas Young that light should be regarded as a wave motion. In fact, life is not so simple! We will now describe the results of the double-slit experiment performed with electrons, but similar results would be obtained if the experiment were repeated with light.

WITH ELECTRONS

Source: an electron 'gun', consisting of a heated wire to 'boil off' electrons from the metal, together with an electric potential to accelerate them.
Screen: a thin metal plate with two very narrow slits in it.
Detector: a screen coated with a chemical 'phosphor' that produces a flash of light every time an electron arrives at it.
Results: Flashes of light signal the arrival of electrons at the detector. As for bullets, we find that electrons arrive singly, in individual 'lumps' of the same size, only at a single place at any one time. If we turn down the intensity of the electron gun, by boiling off fewer electrons per minute, we still see the same size flashes at the detector but with fewer electrons arriving per minute. Again, exactly as for bullets, we can count up the number of flashes we see at any given position of the detector during a given interval of time. As for bullets, this allows us to measure how the probability of arrival of electrons varies as we move along the detector. The magic of quantum mechanics is now revealed. The pattern we see in fig. 1.11 is the interference pattern characteristic of waves, although, as we have said, the electrons always arrive like bullets! This is indeed very strange but things become even more mystifying as we look at this result in more detail.

Let us look at a place where the detector observes a dip or 'minimum' of the interference pattern with both slits open. At such positions we find fewer electrons than would be the case if we repeated the experiment with one slit closed! If we did such a 'one slit closed' experiment with electrons, we would then see the patterns shown in fig. 1.11, exactly as for waves. But if electrons arrive like bullets, how can this be? Does the electron somehow split up into two and half go through each slit? No! Electrons are never seen in halves – just like bullets they are either all there or not at all. Since the invention of quantum mechanics, many people have puzzled to try to find a way out of this dilemma. As far as we know, there is none. It is as if the electrons start as particles at the electron gun, and finish as particles when they arrive at the detector, but the arrival pattern of electrons observed at the detector is as if

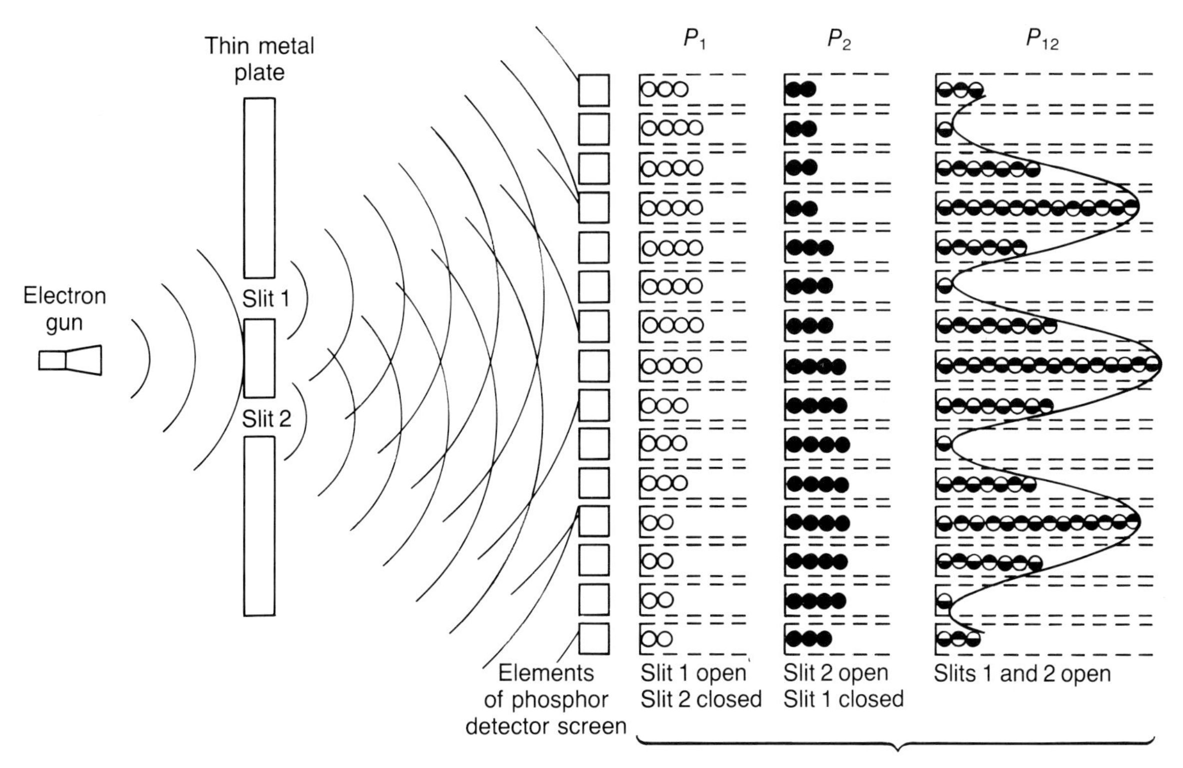

Fig. 1.11 A diagram of a double-slit experiment with electrons. Electrons always arrive with a flash at the phosphor detector at one point, in the same way that bullets always end up in just one of the detector boxes rather than the energy being spread out, as in a wave. The column marked P_1 *shows the pattern obtained with only slit 1 open. Electrons that have gone through slit 1 are represented as open circles, like the bullets of fig. 1.7. Column* P_2 *shows the same thing with only slit 2 open and the electrons that have gone through slit 2 indicated by black circles. These two patterns are exactly the same as those obtained with bullets. The difference lies in the column headed* P_{12}*, which shows the pattern obtained for electrons when both slits are open. This is just the interference pattern obtained with water waves and requires some kind of wave motion arising from each slit as indicated on the figure. It is not the sum of* P_1 *and* P_2 *and so we cannot say which slit any electron goes through. We have indicated this lack of knowledge by drawing the electrons, which still arrive like bullets, as half white and half black circles. This fact, that quantum objects such as electrons possess attributes of both wave and particle motion but behave like neither, is the central mystery of quantum mechanics.*

they travelled like waves in between!

We have seen that the mathematics of the interference curve can be summarized in a very simple equation. We also saw, in the case of water waves, that the interference arose from adding the wave heights or 'amplitudes' for waves from the source to go via slit 1 and via slit 2. The intensity or energy of the wave was then related to the square of the sum of these amplitudes. The same mathematics must hold for the electron interference pattern. In the case of electrons, however, we are not measuring the intensity of a real wave motion but rather the probability of arrival of electrons. From the simple mathematics of the interference curve, moreover, we see that there must be something like the height of a wave in the case of electrons. But what is the meaning of the 'height' of an electron wave? Since the square of this 'height' must give the corresponding

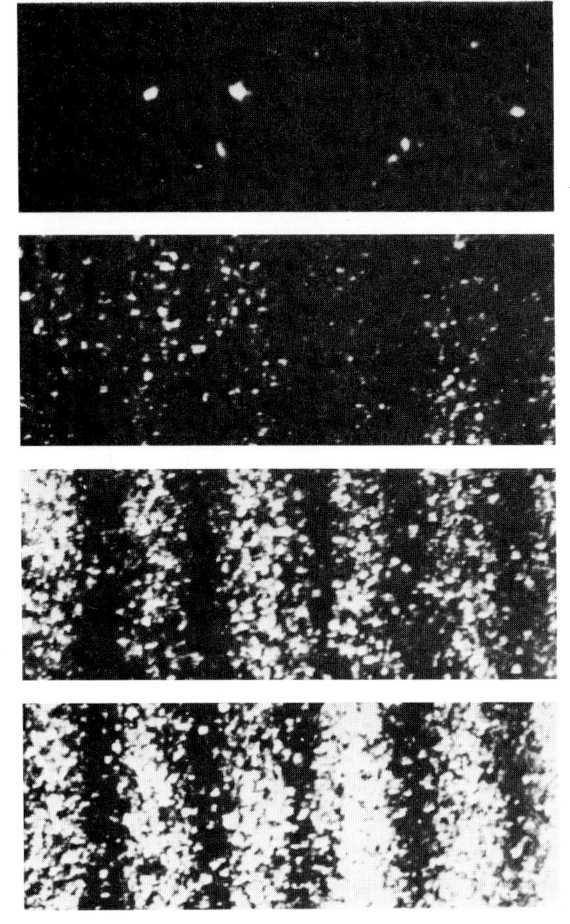

Fig. 1.12 More details of a double-slit experiment performed with electrons. Although interference patterns were once thought of as evidence for wave motion, when looked at in detail it can be seen that the electrons arrive in individual lumps. The top photograph shows a short exposure in which so few electrons have arrived that an almost random pattern of hits is seen. The pictures below show what happens as the exposure becomes longer: more and more electrons arrive until eventually the familiar interference pattern becomes visible.

probability, it is called a 'quantum probability amplitude'. We shall denote such quantum 'heights' or amplitudes by the symbol a. Thus, our equations for the probability of arrival of electrons will all have exactly the same form as for water waves, except that we shall use the symbol P for probability, instead of I for intensity, and a for quantum amplitude, instead of h for height. Thus, the equations for the probability of arrival of electrons with 'both slits open' and with 'one slit closed' take the form

$$P_{12} = (a_1 + a_2)^2$$

$$P_1 = a_1^2$$

$$P_2 = a_2^2$$

and, as before, P_{12} is not equal to the sum of P_1 and P_2:

$$P_{12} \neq P_1 + P_2$$

We must therefore conclude that electrons show wave-like interference in their arrival pattern despite the fact that they arrive in lumps just like bullets. It is in this sense that we can say that quantum objects sometimes behave like a wave and sometimes behave like a particle. You may find this all rather mysterious. It is! We cannot do more to explain the magic of quantum mechanics – all we can do is describe the way quantum things behave. This description is quantum mechanics.

2

Heisenberg and uncertainty

A philosopher once said 'It is necessary for the very existence of science that the same conditions always produce the same results'. Well, they don't!

Richard Feynman

Watching electrons

We have seen that quantum mechanics does not allow us the comfort of being able to visualize the motion of a quantum particle. In a normal game of billiards we can imagine the paths taken by the individual balls (Fig 1.1). Fig. 2.1 shows the physicist George Gamow's attempt to give some impression of how the same game might look if played with quantum particles. Besides illustrating that the notion of a path is no longer valid in quantum mechanics, this cartoon also illustrates another significant difference between the quantum and classical worlds: the exact position of the white ball is not known. Uncertainty has entered physics and replaced the arrogant determinism of Newtonian mechanics.

By the 19th century, physicists had been able to explain vast amounts of experimental observations on objects as different as planets and billiard balls. If an observation differed from the predictions of classical physics, they looked for something they had overlooked to explain the deviation. In 1864, physicists' confidence in the whole edifice of classical physics seemed to be spectacularly verified by an analysis of some irregularities in the orbit of Uranus. These were attributed to the existence of a then undiscovered planet – the subsequent discovery of Neptune was indeed a triumph for Newtonian physics. Thus, by the turn of the 20th century, it seemed that all of physics followed from Newton's laws. If one was given a box containing a certain number of particles, all one had to do to be able to predict the motions of every particle any time in the future (or in the past for that matter) was to measure the present positions and speeds of all the particles. By measuring the speeds and positions sufficiently accurately, these predictions could be made as precise as required. This was the deterministic view of Nature

Fig. 2.1 In this illustration George Gamow has Mr Tompkins playing billiards with quantum billiard balls. The original caption is 'The white ball went in all directions!' In such a world, quantum uncertainty would be a familiar experience.

Fig. 2.2 The existence of the planet Neptune was predicted by Adams in England and by Le Verrier in France at about the same time. After Neptune's discovery there was a familiar Anglo–French squabble about priority, and this French cartoon depicts Adams stealing Le Verrier's results. Le Verrier tried to repeat his success with Neptune by using an anomaly in the orbit of Mercury to predict the existence of another planet, Vulcan, closer to the Sun. This was, of course, not the correct interpretation of the anomaly, which is now understood using Einstein's theory of general relativity.

encouraged by the success of classical physics. The caveat about 'sufficiently accurately' hardly seemed necessary. After all, it was 'obvious' that one could measure anything with essentially no limit to the accuracy of the measurement – all one required was a sufficiently sensitive measuring device.

Quantum mechanics does away with this deterministic view of the future once and for all, and an essential element of uncertainty enters the predictions of physics. How does this come about? It arises because the seemingly innocuous belief of the classical physicists, that they could measure both the position and the velocity of a particle as accurately as they wished, is wrong! In quantum mechanics there is a fundamental limit to the accuracy we can achieve, no matter how ingenious or sensitive we make our measuring devices.

To illustrate this point let us return to the double-slit experiment once more. Remember that we talked in terms of the probability of where an electron would hit the screen. This is because we cannot say with certainty where any particular electron will land. We can only predict the relative

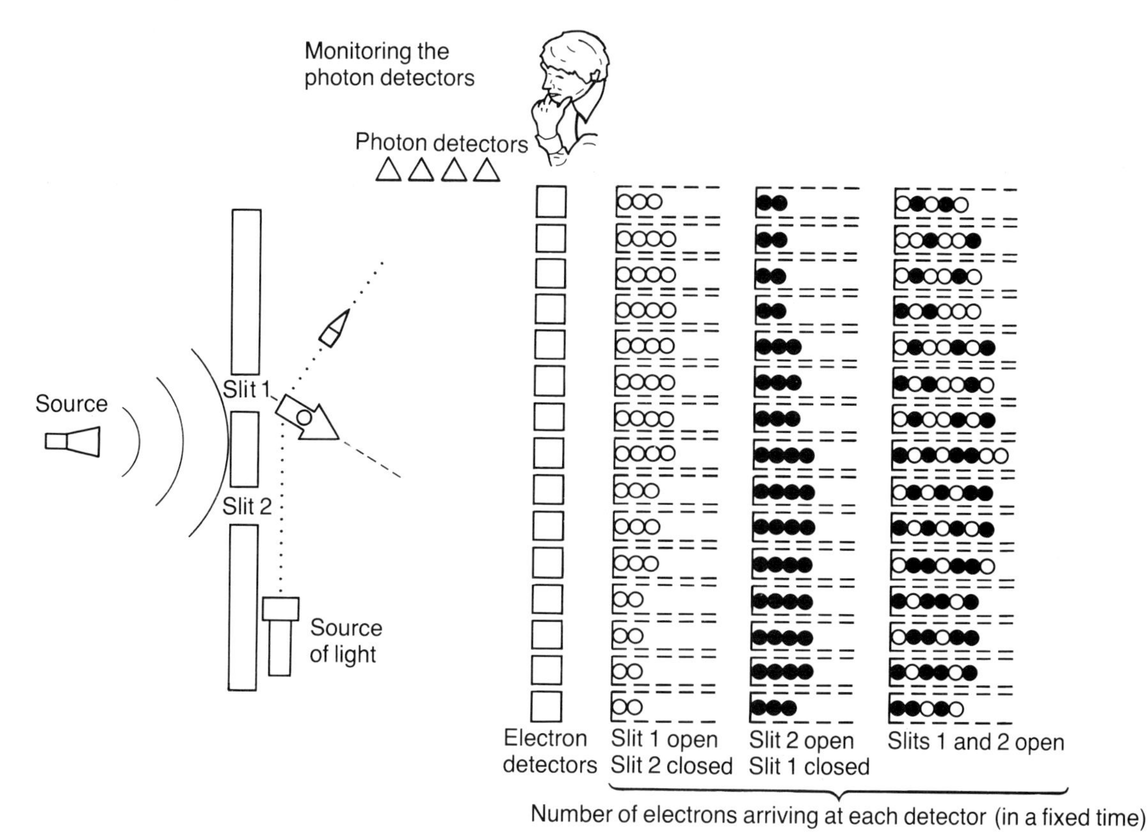

Fig. 2.3 Sketch of the experimental set-up required to observe through which slit the electron passes in a double-slit experiment. Light, in the form of photons, is directed at the slits. In the figure a photon, represented as a small bullet, has hit an electron behind slit 1. The electron is disturbed slightly in its motion and the scattered photon is observed at the photon detectors. The electron patterns obtained with only one of the slits open are almost the same as before, when we did not observe the electron behind the slits. The surprise occurs with both slits open: there is no interference pattern. The small nudges given to the electrons in their collisions with the photons are always sufficient to wash out the interference pattern completely! We can now say with certainty through which slit the electron went but now the electrons are behaving just like bullets. The observed pattern is just the sum of the patterns for slit 1 and slit 2 separately.

chance of it landing at any particular position on the screen.

Now recall the experiment with bullets. This was also described in terms of probabilities. But there is a crucial difference between bullets and electrons. In the case of bullets the probability description was used because of our ignorance of the exact initial direction of the bullet – because of the wobbly gun. However, we could, if we wanted, make a video of the firing of an individual bullet and then watch the bullet's trajectory to the screen on a slow-motion playback. Even if we only saw a part of the bullet's path, that would be sufficient, according to Newton, to determine the rest of the path. Obviously, the bullet must pass through one of the slits, and we can determine which one by looking at the film.

Why can't we do the same sort of thing for electrons? Let us imagine how we would go about trying to establish which slit the electron passes through. To see the electron just after it has passed through one of the slits we must shine some light on it and observe the reflected light. Let us therefore modify the experimental apparatus by inserting a light source behind the slits. We now arrange things so that if an electron passes through slit 1 we see a flash behind slit 1, and similarly for slit 2. If we now do the experiment what do we see? Well, the first important result is that we never see a sort of half-flash behind both slits simultaneously. There is always a whole flash behind either slit 1 or slit 2. We can therefore now divide up the electrons arriving at the detector into two groups, according to whether they went through slit 1 or slit 2. What's all this quantum nonsense then? The electron obviously either goes through slit 1 or slit 2. So indeed it does, when we watch the electron. But, if we now look at the arrival pattern of electrons at the screen, we see no interference pattern! The result is just the same as we obtained with bullets!

Amazingly, we have a different result depending on whether or not we switch on the light in order to watch the electrons! The resolution of this apparent paradox lies in the quantum nature of light itself. Remember our discussion of the photo-electric effect in chapter 1. When light interacts with matter, it displays its particle-like character. Light, like electrons, arrives in definite chunks of energy called photons. To see an object, therefore, we must bounce at least one photon off it. Now comes the crux of the argument. When we shine light on a bullet, its motion is not noticeably disturbed because the amount of energy in an individual photon is tiny compared with that of the bullet. Electrons, on the other hand, are very delicate quantum objects. Shining light on electrons gives them a jolt that disturbs their motion significantly. A more detailed analysis reveals that this disturbance is always just enough to wash out the interference pattern!

You may think we can turn the light down very low and make the disturbance so small that the interference pattern is not destroyed. This idea ignores the way light works. If we reduce the light intensity, we merely cut down the number of photons emitted per second. Now that we only have a few photons around, there is a good chance that an electron can sneak past without being seen. We must therefore make a third category for electrons arriving at the screen. These are the ones that we missed and that we cannot say went definitely either through slit 1 or through slit 2. If we look at the arrival pattern of these 'missed' electrons we see the interference pattern once more!

This is what Feynman calls the 'logical tightrope' of quantum mechanical thinking. If we have an experiment which can detect which slit the electron goes through, then one can say with certainty that the electron does go through one or other of the slits. If, however, we have no way of telling which slit the electron went through, then we may not say that the electron goes through either one slit or the other!

Werner Heisenberg (1901–76) was in his early 20s when he performed his fundamental work on quantum theory. He was awarded the Nobel Prize in 1932 for his discovery of the uncertainty principle.

Heisenberg's uncertainty principle

It is clear that quantum mechanics is a very cunning and subtle theory. For the double-slit experiment, we have seen that establishing which slit the electron went through destroys the interference pattern. This result is indicative of a very general principle of quantum physics, now named after its discoverer, Werner Heisenberg. It was Heisenberg who first pointed out that the new laws of quantum mechanics imply a fundamental limitation to the accuracy of experimental measurements. In our everyday world we can certainly imagine making measurements sufficiently delicately so that the act of measurement does not cause a perceptible disturbance. In the quantum world this is not the case. Light energy arrives in lumps, and making a measurement necessarily gives a significant jolt to the object on which we are making the measurements. Furthermore, there is no way that we can reduce the jolt to zero, even in principle. For objects of microscopic dimensions such jolts are not negligible. This is the essence of Heisenberg's uncertainty principle.

The uncertainty principle can be written down in a precise mathematical form as follows. In our discussion of the deterministic nature of classical physics, we imagined measuring the position and velocity of every particle in a box. We shall often refer to this collection of particles and surrounding box as a 'system' and talk about making measurements on 'the system'. Physicists also usually denote a measurement of the position of a particle by the symbol x, but instead of speed or velocity, they prefer to talk about a quantity called 'momentum'. Momentum is just the mass of the particle multiplied by its velocity, and it is a familiar concept in everyday life. A car moving with a speed of 10 miles per hour has more momentum than a football moving with the same speed (and consequently the car will do more damage when it collides with something!). Physicists usually represent momentum by the symbol p. Now in making measurements on a quantum system, it is not possible to measure the quantities x and p as accurately as we would wish. There is always some minimum error or uncertainty Δx and Δp, associated with their measurement. What Heisenberg discovered, and what actually keeps quantum mechanics from being internally inconsistent, is that the uncertainty in the position measurement, Δx, and the uncertainty in momentum, Δp, are inextricably linked together. If you want to measure the position of a particle very accurately, you inevitably end up by disturbing the system rather a lot, and, consequently, introducing a large uncertainty in the momentum of the particle. Why is this? Well, to determine the position very accurately it is necessary to use light with a very short wavelength, since the wavelength of the light determines the minimum distance within which we can locate the particle. Very short wavelength light is light of very high frequency. Here comes catch 22. The energy of a photon depends on its frequency according to a formula first guessed by Max Planck. The formula is very simple: it says that the photon's energy is proportional to its frequency. We can write down this formula relating the energy of the photon, E, to the frequency, f, of the light as follows:

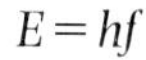

$$E = hf$$

photon energy E = Planck's constant h times frequency f

Max Planck (1858–1947) around 1900. His radical solution to the problem of radiation from hot bodies was the first introduction of quantum ideas. The importance of this work was recognized by the award of the Nobel Prize in 1918, but Planck remained unhappy with the quantum revolution his work had ushered in.

The constant of proportionality, h, is known as Planck's constant. Armed with this result we can now return to the problem of making an accurate measurement of position. We see that in order to locate the particle very precisely we must use high frequency light with a large value of f. But, such high frequency light will arrive in photons with a very large energy which therefore give the quantum system a very large kick. Similarly, if we want to know the momentum very accurately, we must give the system a very small kick. According to Planck's formula this means using light of low frequency. Low frequency means long wavelength and this in turn means a large uncertainty in the measurement of position!

Heisenberg's uncertainty principle relates the uncertainties in position and momentum measurements in the following way:

$$(\Delta x)\,(\Delta p) \approx h$$

$$\begin{pmatrix}\text{uncertainty}\\ \text{in position}\end{pmatrix} \text{ times } \begin{pmatrix}\text{uncertainty}\\ \text{in momentum}\end{pmatrix} \text{ is approximately equal to } \begin{pmatrix}\text{Planck's}\\ \text{constant}\end{pmatrix}$$

This equation puts into mathematical form the correspondence we discussed above. If you want to make the uncertainty in position, Δx, very small, then the uncertainty in momentum, Δp, cannot also be small. If both were small, the product of Δx and Δp would not satisfy Heisenberg's equation, which says that the product of the uncertainties must always be

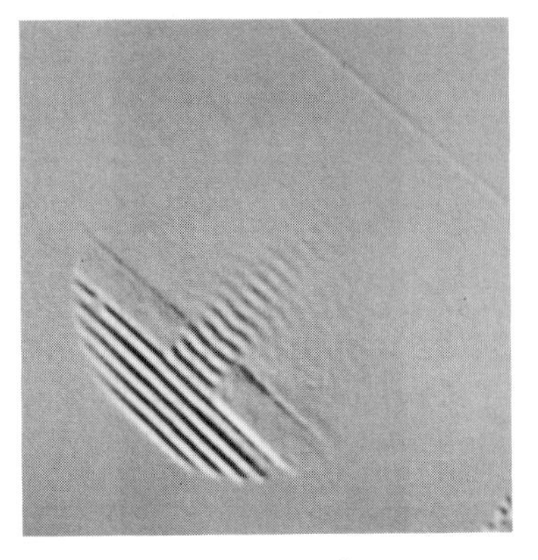
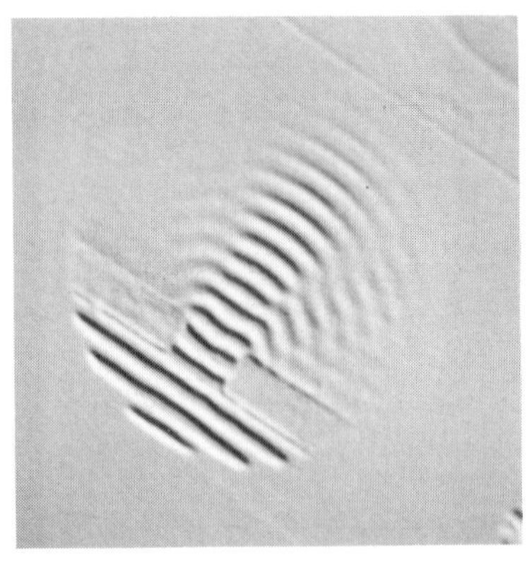

Fig. 2.4 This sequence of photographs shows how waves spread out as they pass through a slit. This spreading is known as 'diffraction' and is largest for long wavelengths. This effect limits the amount of detail visible in optical instruments.

approximately equal to Planck's constant h. Notice that this is the second equation of quantum mechanics that we have come across that contains this mysterious constant. Planck's constant, h, may be measured experimentally in experiments on the photo-electric effect. Its value turns out to be so tiny that Heisenberg's restrictions on the accuracy of measurements have a negligible impact on our everyday observations of cars or billiard balls. The smallness of h is of course the reason it was so long before physicists were able to observe any of the strange quantum effects that we shall describe in the subsequent chapters of this book.

There is an amusing postscript to the story of Heisenberg and his uncertainty principle. A few years before his work on the uncertainty principle, Heisenberg was finishing his doctoral thesis in Munich under the supervision of a famous theoretical physicist, Arnold Sommerfeld. At his oral examination, Heisenberg antagonized one of his examiners, a very eminent professor of experimental physics, Wilhelm Wien, by his inability to answer questions of a fairly elementary nature concerning the resolving power of optical instruments. As a result, it was only after special pleading by Sommerfeld that Wien was persuaded to pass Heisenberg, and then only with the lowest possible passmark. A few years later, Heisenberg's ignorance of such a basic point in classical optics caught him out. He illustrated his new uncertainty principle by considering a 'gamma-ray microscope' – a hypothetical microscope that could be used to look at electrons with very short wavelength gamma-ray light (see appendix 1). Unfortunately, Heisenberg forgot the lesson of his uncomfortable oral examination and his analysis of this problem did not take the resolving power of the microscope into account! It was left to another great man of physics, Niels Bohr, to point this out gently to Heisenberg, and close this loophole in his argument.

Uncertainty and photography

The probabilistic or statistical nature of quantum processes can be seen not only with electrons but also with light. Imagine looking at a faint star on a dark night. We see the star because light from the star causes chemical changes in the retinal cells of the eye. In order that these chemical reactions take place it is necessary that the light energy arrives in localized chunks – photons. The eye is quite a good light detector: a single photon can stimulate a retinal cell. In general, however, many photons are absorbed by the eye without reaching a light sensitive cell. For this reason only a few photons in every hundred or so that enter the eye are detected. Obviously the chemical changes involved in seeing something must be reversible – in fact, the cell reverts to its normal state after about one-tenth of a second. It is this short light storage period that limits the sensitivity of the eye for detecting faint objects. Photography can overcome this limitation of the eye by storing the

Fig. 2.5 Four photographs of the Andromeda Galaxy showing how the amount of detail visible increases with increasing exposure time.

changes in a permanent way on photographic emulsion.

In the same way as for the eye, single photons can cause chemical changes in the specially prepared photographic material in a roll of film. What is the active ingredient in the photographic emulsion? If you know this, you can answer one of the questions in the question-and-answer game *Trivial Pursuit*. The question is 'What company is the world's largest user of silver?' and the answer is 'Kodak'. Photographic film consists of lots of individual grains of a silver compound, in which the silver atoms are 'ionized'. A silver ion is an atom of silver which has lost one of its negatively charged electrons. Normally, in a neutral atom, the total charge of all the electrons exactly cancels the positive charge of the nucleus. A silver ion, therefore, has a net positive charge. When a photon is absorbed by the emulsion, an electron is sometimes emitted, in the same way as electrons are knocked out of a metal in the photo-electric effect. This electron can now be attracted by a silver ion to form a neutral atom of silver. Left to itself, the

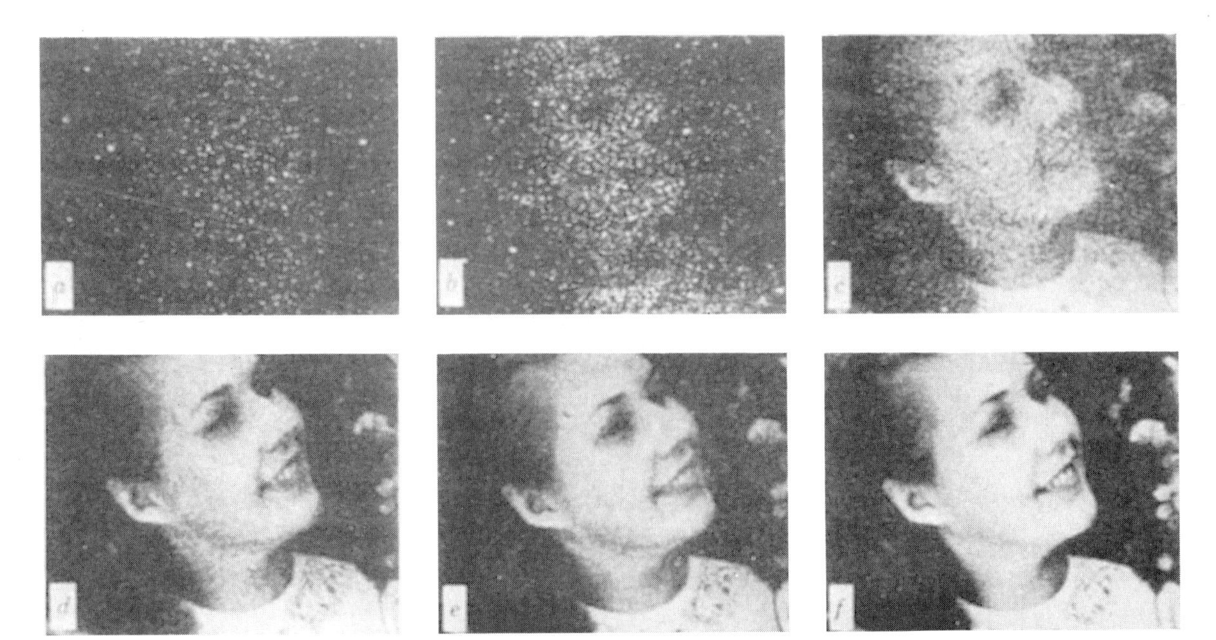

Fig. 2.6 This sequence of photographs of a girl's face shows that photography is a quantum process. The probabilistic nature of quantum effects is evident from the first photographs in which the numbers of photons are very small. As the number of photons increases the photograph becomes more and more distinct until the optimum exposure is reached. The number of photons involved in these photographs ranges from about 3000 in the lowest exposure to about 30 000 000 in the final exposure.

neutral silver atom, surrounded by the ionic silver compound, is unstable, and will eventually eject the electron and revert back to an ion. However, if, before this happens, other photons have produced several other neutral silver atoms nearby, a stable 'development centre' consisting of a small number of atoms can be formed. In contrast, each grain of the emulsion contains billions of silver ions. However, when the film is developed, this tiny group of neutral silver atoms induces all the remaining silver ions in the grain to be deposited as opaque metallic silver. How does photography help us to see very faint stars? Well, for such very faint objects the chance of forming a development centre is rather low, because of the small number of photons reaching the Earth from the star. But, if we wait longer and increase the exposure of the photographic plate, the chance of this happening will increase. The photographs shown in fig. 2.5 are of the Andromeda Galaxy taken at different exposures. The detail of the outer spiral arms is invisible to the eye but is revealed in the long exposure photographs.

Now let us consider taking an ordinary photograph with a camera. Fig. 2.6 shows several photographs of the same person taken at different exposures. In the top left picture about 3000 photons enter the camera. Most of these photons are absorbed without causing a permanent change in the emulsion. It is evident that 3000 photons are not enough to give us a recognizable image and the photograph appears like a more-or-less random series of dots. However, as we increase the exposure the number of photons entering the camera increases. The top right picture involves about 10 000 photons and already, although there is no clear image, a blurred impression of an image is beginning to show up. The improvement continues as we increase the number of photons, and the final exposure involves more than 30 000 000 photons. In this last picture, the image intensity seems to vary smoothly from place to place on the photograph, whereas in fact we know it is built up out of individual development centres created by the arrival of individual photons. Furthermore, although in the lowest exposure photograph the positions of the bright dots, signifying the presence of a development centre in that grain of emulsion, seem pretty random, we see that they are not. Centres are more likely to develop in places where the image will eventually be bright. Thus, even in the commonplace action of taking a photograph, we can see the quantum mechanical, probabilistic

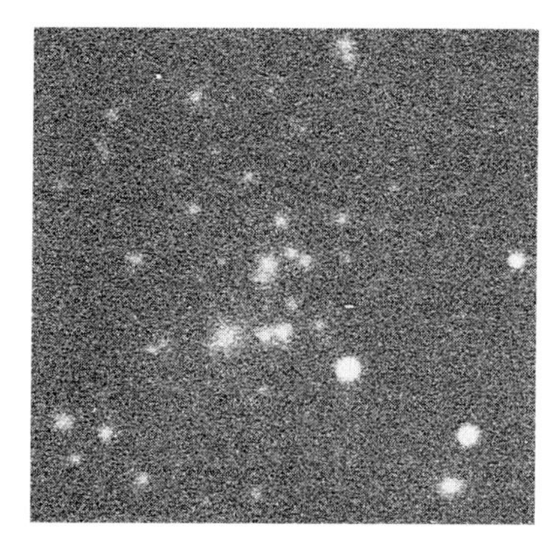

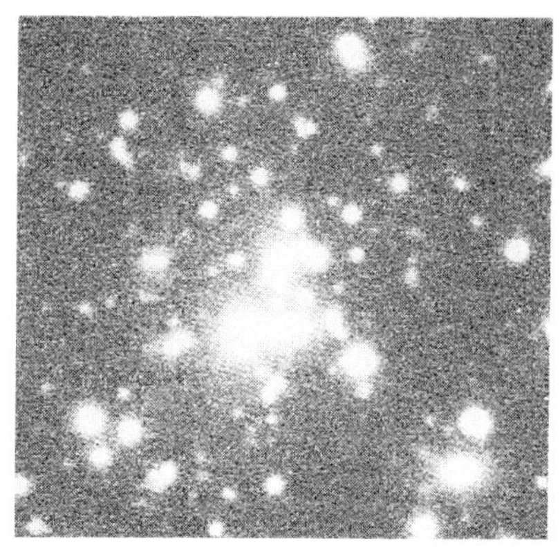

Fig. 2.7 A comparison of a photograph and a CCD image of a cluster of galaxies. The left hand photograph is taken in a 90 minute exposure with the 100 inch telescope on Mount Wilson, California. The right hand picture shows a CCD image of the same region using the 60 inch telescope at Mount Palomar, also in California, with an exposure of 25 minutes. This illustrates the dramatic improvement in sensitivity that can be achieved using CCD devices.

nature of light. We cannot predict with certainty where any particular photon will land, or in which grain a development centre will be produced. All we can talk about are probabilities.

As we have seen, photographic emulsions are not sensitive to individual photons – several neutral atoms must be produced to form a development centre. Nowadays in astronomy, a new type of detector is replacing photographic plates. This is the so-called 'charge coupled device', or 'CCD', which can detect the arrival of a single photon. It is much more efficient than photography for the detection of very faint stellar objects, as can be seen in fig. 2.7. A CCD consists of an array of small 'photon detectors' laid out on a silicon chip. As we shall see in a later chapter, silicon is an example of a class of materials that are called 'semiconductors'. Roughly speaking, semiconductors are substances whose electrical properties are halfway between metals, which allow electric current to flow easily, and insulators, which do not allow currents to flow at all. Silicon also has the property of requiring very little energy to release electrons from their parent atoms. By carefully adjusting the temperature at which the CCD is operated, the silicon can be made sensitive to the passage of a single photon. Each 'detector' is in fact just a small region of silicon where the electrons liberated by the arrival of photons can be collected. A measurement of the accumulated charge at each position over the array then corresponds to the pattern of photons striking the CCD.

Richard Feynman was born in 1918 just outside New York. He has made many contributions to many areas of theoretical physics. His 'sum over histories' way of looking at quantum amplitudes now plays a central role in modern quantum field theory. During the war, he worked on the Los Alamos 'Manhattan' project that led to the development of the first atom bomb. One of the great men of quantum mechanics, Niels Bohr, picked on Feynman to try out his new ideas because he was the only person at Los Alamos who would not be in awe of his reputation and would tell him if his ideas were lousy!

We have now seen how quantum uncertainty manifests itself in something as simple as taking a photograph. Feynman suggested yet another way of looking at this quantum uncertainty. This is in terms of 'classical' and 'quantum' paths for the particle, and this insight has turned out to be of great importance in modern quantum theory.

Feynman's quantum paths

There is another interesting way in which to view the similarities and differences between classical and quantum physics. Consider yet again the double-slit experiment, and suppose that we wish to calculate the probability for an electron to leave the source (S) and arrive at some position at the detector (D). In order to calculate the observed arrival pattern we found that we had to add the probability amplitudes for the paths 1 and 2

$$a = a_1 + a_2$$

to obtain the complete quantum amplitude. The probability of arrival at any point is then obtained by squaring this amplitude:

$$P = (a_1 + a_2)^2$$

This is the quantum mechanical recipe that is required to account for the

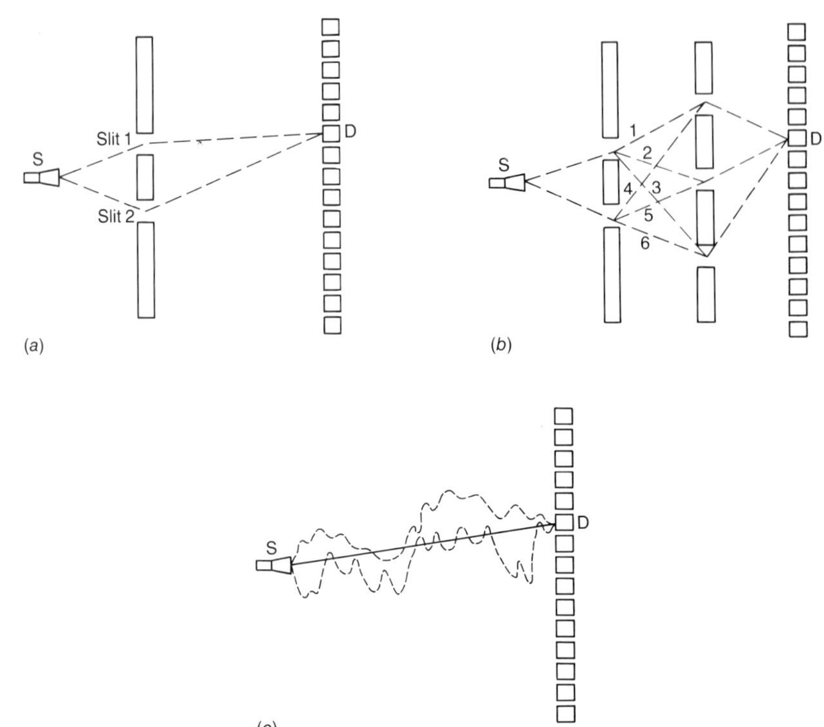

Fig. 2.8 The quantum amplitude may be obtained by adding the amplitudes for all possible paths between source S and detector D. (a) The original double-slit experiment with two possible paths for the electron. (b) With two screens between source and detector and a total of five slits there are now six possible paths. (c) Adding more screens and then cutting more and more slits leads to the situation where there are no screens at all! The quantum amplitude for an electron to travel from S to D may therefore be regarded as a sum over all possible paths. Two of the infinite number of possible quantum paths are shown as dotted lines in this figure. The path taken by a classical particle is shown as a solid line.

interference pattern observed by experiment. We will return to the question of exactly how we determine the appropriate amplitudes for each path when we discuss Schrödinger's equation in the next chapter.

For the moment let us just accept this rule and consider what happens if we complicate the experiment by introducing a second screen, with three more slits, as shown in fig. 2.8*b*. There are now six possible paths from S to D, and, according to our quantum mechanical rule, we must add up the amplitudes for all these paths to obtain the total probability amplitude:

$$a = a_1 + a_2 + a_3 + a_4 + a_5 + a_6$$

total amplitude = sum of amplitudes for each possible path

The probability of arrival is again just the square of this. Now imagine what happens if we put more and more screens with more and more slits between the source and the detector. To obtain the total probability amplitude we must still add up all the amplitudes for all the many possible paths. If we carry on putting in screens, eventually we will fill up the space between S and D with screens. If we now make more and more slits in each screen, in the end we will have no screens at all! This line of reasoning led Feynman to write down an expression for the total probability amplitude to go from S to D, in the absence of any screens or slits, as a sum over amplitudes for all possible paths between S and D. In fig. 2.8*c* we have indicated two such possible 'quantum paths', together with the straight line trajectory that a bullet would follow in going from S to D in the absence of any screens. In classical physics there is only one possible path, whereas in quantum physics we must consider all possible paths between S and D to obtain the correct probability of arrival.

We can also see a connection between summing over all quantum paths and the quantum uncertainty principle. Let us consider first an example of classical motion. We show a section of a 'roller coaster' in fig. 2.9. If the carriage is placed on the roller coaster at its lowest point, then, according to classical physics, it will remain at rest indefinitely unless we do something about it. Below the picture of the roller coaster we have represented this state of affairs by a graph of the position of the carriage (along the horizontal axis) against time (in the vertical direction). The straight vertical line is just telling us that the carriage never moves.

What happens with quantum objects such as electrons? As we describe in detail in the next chapter, it is possible to set up an arrangement of electric fields that act in the same way for the electron as the roller coaster track does for the carriage. But, according to quantum mechanics, an electron is not allowed simply to sit at rest in the bottom of the valley! If it was allowed to do

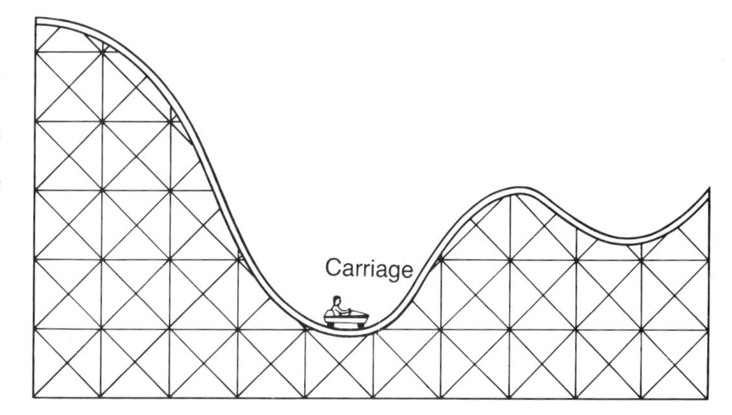

Fig. 2.9 A roller coaster with the carriage at rest in one of the valleys. Heisenberg's uncertainty principle does not allow a quantum roller coaster to remain at rest: instead the quantum carriage would be forever jiggling back and forth across the bottom of the valley.

so, we would know both the position and the momentum of the electron simultaneously, and Heisenberg's uncertainty principle tells us that this is not possible. What happens then? According to quantum mechanics, the electron must be constantly jiggling around near the bottom of the valley and can never be at rest. What would a graph of the position of the electron against time look like? Obviously it will not be just the simple vertical line of the roller coaster carriage. Instead, it will be some sort of complicated, jagged curve corresponding to all the quantum jiggling. In fact, using Feynman's 'sum-over-paths' way of looking at quantum mechanics, it is possible to generate some typical 'quantum paths' for such an electron in a computer simulation. Some examples of such computer-generated paths are shown in fig. 2.10.

A MATHEMATICAL CURIOSITY

At this point we shall make a diversion from our main theme to include a brief discussion of some rather curious mathematical curves.

The quantum paths shown in fig. 2.10 represent 'snapshots' of the position of the electron over a given period of time. Going from (*a*) to (*b*) to (*c*), we have divided up the same period of time into finer and finer intervals. This is like increasing the magnification at which we look at the curves of the electron's motion. As can be seen, these quantum paths look very jiggly no matter what magnification we choose to look at them. This 'looking the same at all length scales' is a property that is characteristic of an interesting type of curve studied by mathematicians. We are accustomed to the idea that a line has a definite length associated with it. Thus, we can measure the length of a running track in metres, for example. Similarly, an area is associated with 'length squared'. A football pitch has an area measured in 'square metres', for example. We may therefore say that a line or an area has a magnitude expressed in terms of a 'length to the power *D*', where *D*, the 'dimension' is one for a line and two for an area. The curves we have here,

Fig. 2.10 Typical paths for a quantum roller coaster. The jagged curves connect the positions of the carriage as seen after equal time intervals. The sequence of photographs shows the effect of looking at the motion more frequently. As we sample the motion at finer and finer time intervals the path becomes more and more jiggly. This property of being jiggly at all time scales is the characteristic property of a 'fractal'.

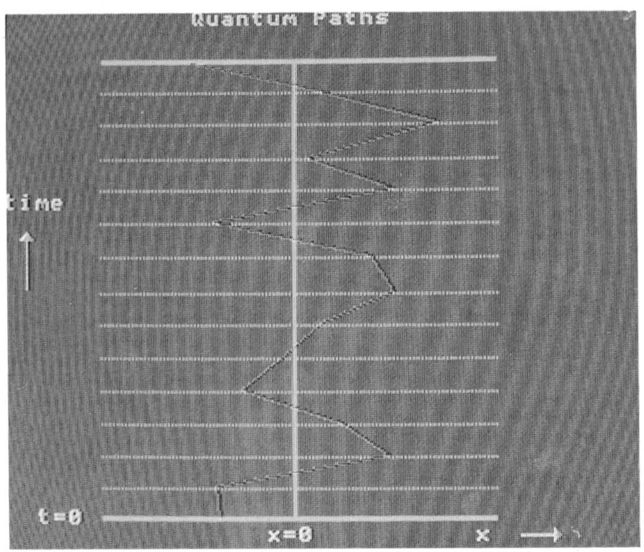

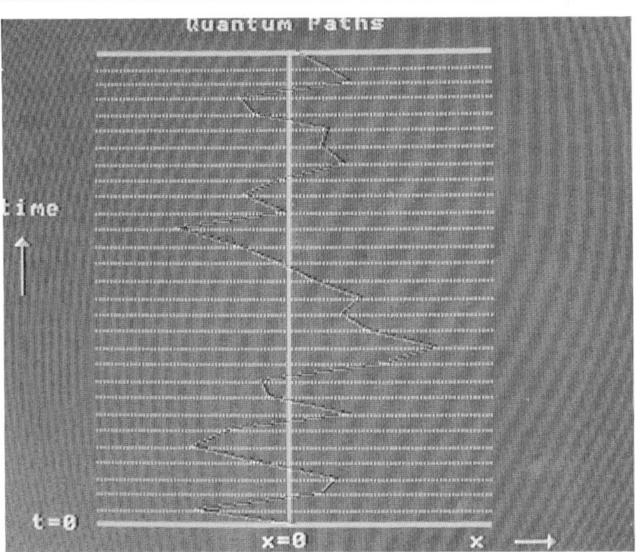

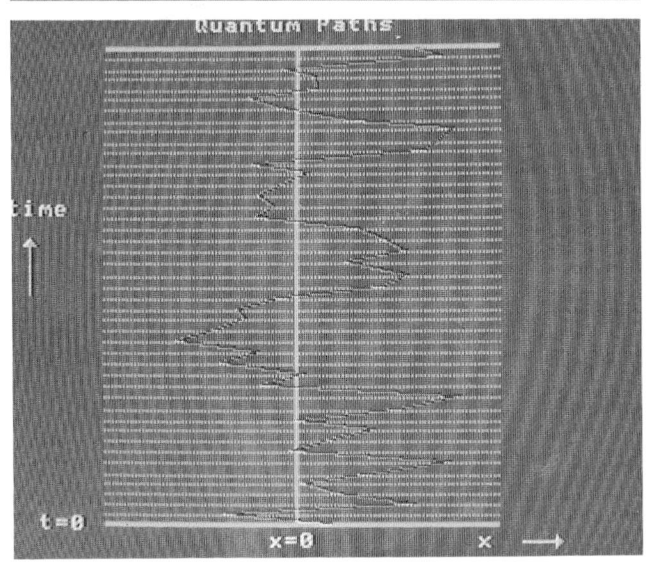

however, turn out to be so irregular and jiggly that they 'fill more space' than an ordinary line. Such curves are called 'fractals' and may have a dimension that is greater than one!

A much-loved example that gives some idea of the curious nature of fractals is that of Lewis Richardson's measurement of the length of the coastline of Great Britain. Consider measuring the length of a straight line using a pair of dividers set open so that the points are a certain distance apart, to pace off 'steps'. It is certainly plausible that the value we obtain for the length of the line will not depend on how far apart we have set the dividers. This is not true for a measurement of the length of the coastline of Great Britain. If we use a large spacing of the dividers, and go round the coast on a fine scale map, using the dividers as carefully as we can, it is as if we were using a much coarser scale map. The setting of the dividers is too big to follow all the smaller inlets and headlands. We can only follow these details if we use a smaller setting for the dividers. Obviously, the length of coastline between two points is longer than the straight line distance between them, but, equally obviously, the distance we measure will be larger and larger the finer the scale on which we use to measure it. The coastline looks jiggly whatever the scale we choose, and our answer for the length will increase as we look in finer and finer detail. This effect may be part of the reason why different encyclopedias give different values for the length of land frontiers between different countries. Spanish and Portuguese encylopedias, for example give distances for the length of the boundary between the two countries that differ by as much as 20%!

It is clearly somewhat unsatisfactory to have a definition of length that depends on the scale at which the measurement is made. Roughly speaking, the fractal dimension of a curve is defined so that there is a 'fractal length' that is independent of the scale of measurement. As we have said, this dimension need not be the same as the ordinary dimension of a curve.

Fig. 2.11 A computer-generated fractal landscape showing an impressively realistic misty scene. Many natural features can be approximated as fractals and such artificial landscapes are now used extensively in modern science fiction movies.

Indeed, from Richardson's data on the west coast of Great Britain, Benoit Mandelbrot, who was the first popularizer of fractals, deduced that the coastline had a fractal dimension of around 1.2! Similarly, the quantum paths shown in fig. 2.10 are fractals with dimension 1.5. Mandelbrot has produced a book on fractals which contains many beautiful computer-generated pictures showing the similarity between natural features, such as snowflakes or clouds, and certain types of fractals. Computer-generation of landscapes using fractals is now one of the standard tricks of the trade of the new breed of science fantasy films. A striking example of a fractal 'lunar landscape' is shown in fig. 2.11.

3

Schrödinger and matter waves

'Where did we get that [equation] from? Nowhere. It is not possible to derive it from anything you know. It came out of the mind of Schrödinger.

Richard Feynman

de Broglie's matter waves

The early struggles of physicists towards a quantum theory were mostly concerned with attempts to understand the nature of light. The traditional picture of light as a wave motion had been challenged by Planck and Einstein. They had shown that certain experimental results, that were impossible to understand in terms of a wave picture, could be easily explained if light was thought of as a stream of particles, now called photons. William Bragg, who, with his son, won the 1915 Nobel Prize for studies of crystal structure using X-rays, summarized this dilemma for physics by exclaiming in despair that he was teaching the corpuscular theory of light on Mondays, Wednesdays and Fridays, and the undulatory theory on Tuesdays, Thursdays and Saturdays! Physicists were still wrestling with these apparently contradictory properties of light when, in 1924, Prince Louis de Broglie (pronounced 'de Broy') suggested that all matter, even objects that we usually think of as particles – such as electrons – should also display wavelike behaviour! This revolutionary idea was completely unexpected and, what is more, was included by de Broglie in his PhD thesis. Like most people, physicists are generally fairly reluctant to accept any wild new idea, especially if there is not a shred of experimental evidence to support it. Predictably, therefore, de Broglie's examining committee in Paris were distinctly unsure as to what to do about the thesis. In fact, one of the examining committee, Professor Langevin, himself an eminent physicist of the day, was described by de Broglie as being 'probablement un peu étonné par la nouveauté de mes idées' ('probably a bit stunned by the novelty of my ideas')! Langevin, in fact, asked de Broglie for another copy of the thesis which he then sent to Einstein for his opinion. Einstein was impressed, and said later of de Broglie's work, 'I believe it is a first feeble ray of light on this worst of our physics enigmas'. Fortunately for the examining committee, they made the right decision and gave de Broglie his doctorate. For it was only a few years later, in 1927, that wavelike behaviour for electrons was convincingly demonstrated – by Davisson and Germer in the USA, and by G. P. Thomson in Scotland – and both de Broglie, in 1929, and Davisson and Thomson, in 1937, received Nobel Prizes for their work on 'matter waves'.

Prince Louis de Broglie was born in 1892 from a noble French family. His great-great-grandfather, in fact, died on the guillotine during the French revolution. de Broglie initially obtained a degree in history, but while serving in the French army during the First World War he became interested in science. He was involved with radio communication and was stationed on top of the Eiffel Tower. His simple mathematical relationship connecting the wave and particle properties of matter earned him the Nobel Prize in 1929.

If all 'particles' can behave like waves, why did it take physicists so long to observe these matter waves? Why don't we see wavelike behaviour for bullets, billiard balls, or even cars? Again, the answer to these questions lies in the smallness of Planck's constant. According to de Broglie, the wavelength of the matter waves of such everyday objects is very tiny. de Broglie suggested that a particle travelling with a certain momentum p has an associated matter wave of wavelength λ given by the expression

$$\lambda = h/p \qquad \text{(de Broglie's relation)}$$

wavelength = Planck's constant divided by momentum

As we saw in our discussion of Heisenberg's uncertainty principle, it is Planck's constant that characterizes the size of all quantum effects. But how

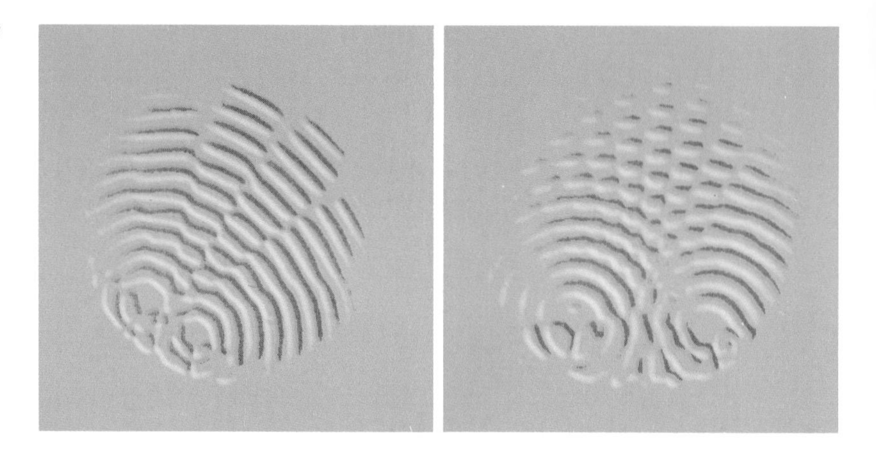

Fig. 3.1 These two photographs show how the interference pattern is altered when the separation of the sources is increased. Larger separations cause the interference bands to move closer.

does its extreme smallness explain why wavelike behaviour for everyday objects is not observed? Well, what we did not say when we talked about the double-slit experiment was that in order to see wavelike interference effects, the separation of the two slits must be about the same size as the wavelength of the objects – photons or electrons – that are doing the interfering. Since the de Broglie wavelength of a bullet fired from a gun is much smaller than even atomic dimensions, it is impossible to devise an experiment that will show interference with bullets, or indeed any other everyday object. If, on the other hand, we could increase the size of Planck's constant things would look very different, as in Mr Tompkin's nightmares!

Fig. 3.2 Another view of Mr Tompkins' Wonderland in which Planck's constant is much larger than in our world. The caption read 'Sir Richard was ready to shoot, when the professor stopped him'. The professor went on to explain 'there is very little chance of hitting an animal when it is moving in a diffraction pattern'.

Schrödinger's equation

At the time he discovered his now famous equation, Erwin Schrödinger was a moderately successful, middle-aged, Austrian physicist working in Zurich. Professor Debye, head of the research group in Zurich, had heard about these peculiar waves of de Broglie and asked Schrödinger to explain these ideas to the rest of the group. Schrödinger did so, and when he had finished Debye remarked that it all seemed rather childish – to deal properly with waves one should have a wave equation to describe how the wave moves from place to place. Stimulated by this remark Schrödinger went away and discovered the equation that now bears his name. This was a vital breakthrough because it enabled physicists to calculate how quantum probability waves move, and therefore make precise predictions which can be compared with experiment. Just as Newton guessed the simple laws that describe all of classical physics, so Schrödinger guessed the law that describes the motion of quantum objects. Before we write down Schrödinger's equation, in order to make it seem less like a rabbit out of a hat, it is helpful to introduce the idea of 'conservation of energy'. This we can do by looking at the motion of familiar objects in our everyday world.

Imagine yourself back on the roller coaster again. If we start from rest at the top of the hill on the left, as we slide down the track the car will go faster and faster until it reaches the bottom. We then start going up the other side, and will gradually slow down until the car comes to a stop. The motion of the car illustrates the principle of conservation of energy. When we start out, high above the valley, we are at rest and have no 'kinetic energy' – energy due to our speed. At the bottom, when we are travelling fastest, we have a lot of kinetic energy. As we travel up the other side, we steadily lose kinetic energy until we are once more at rest. Where has all the kinetic energy gone? As we climb up the slope, we must use up energy in raising the weight of the carriage and its occupants back up to the top of the hill. We describe this by saying that we do work against the pull of gravity and, as we gain height, we are said to gain gravitational 'potential energy'. When we started out, we had no kinetic energy, but because we were high up, we were able to convert our gravitational potential energy into kinetic energy by rolling down the slope and losing height. The total amount of energy is always the same, but the form of it may change. In principle, therefore, the roller coaster will slide down to the bottom of the valley and then coast up the other side to exactly the same height as we started from. In fact, of course, a real roller coaster car will not quite reach the same height as it started from, because some of the initial potential energy will be lost to the surrounding environment in the form of energy used in heating up the tracks, in causing noise and so on. These are all so-called 'frictional' energy losses. To keep things simple we shall ignore such losses and our roller coaster should be imagined as being

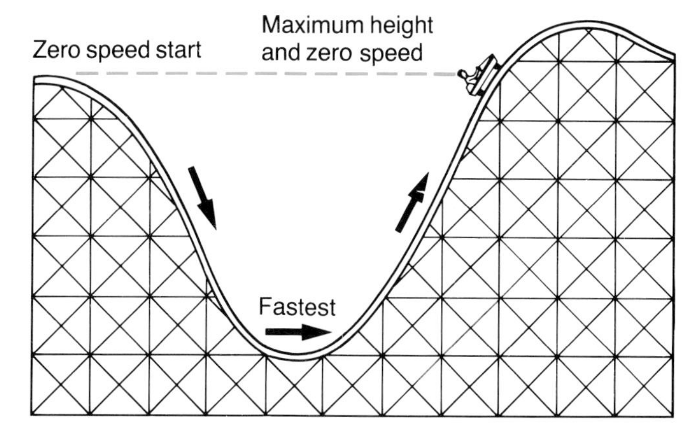

Fig. 3.3 An idealized roller coaster ride illustrating conservation of energy. At the top left hand side the carriage starts to move from rest. In this position it has zero kinetic energy due to motion and maximum gravitational potential energy due to its height. As the carriage rolls down the track the gravitational energy is converted to kinetic energy so that at the bottom it has zero gravitational energy and maximum speed and kinetic energy. The car then slows as it climbs up the hill on the other side converting back kinetic energy into gravitational energy. On an idealized roller coaster with no energy losses due to heating the track, generating noise, and so on, the carriage will climb up to exactly the same height from which it started.

very slippery and shiny. Conservation of energy for our roller coaster example can be summarized by saying that the total energy, for which we use the symbol E, is constant, but can be made up of varying amounts of kinetic energy, K, and potential energy, which is traditionally denoted by V. As an equation this reads

$$E = K + V$$

total energy = kinetic energy + potential energy

This equation is true for any position of the roller coaster car on the track and at all times.

Before we leave this example, there is another way of writing the energy equation that will be useful to us later. As we have said in the preceding chapter, the momentum p of an object is just given by its mass m multiplied by its velocity v:

$$p = mv$$

momentum = mass times velocity

Now, according to Newton's laws the kinetic energy and momentum are related by the following equation:

$$K = p^2/(2m)$$

Thus, our book-keeping relation for energy may also be written as

$$E = p^2/(2m) + V$$

which is now an equation relating total energy, momentum and potential energy.

What has all this to do with electrons and the Schrödinger equation? In the previous chapter we remarked that it was possible to set up a similar 'roller coaster' for electrons. Quantum objects like electrons still obey the principle of conservation of energy – we are not allowed to create or lose energy, even at the quantum level. But, just like our real roller coaster example, energy can be changed from one form to another. In this case, however, the relevant form of potential energy is not that due to gravity, but, rather, electrical potential energy. Electrons, with their negative electric charge, are attracted to a region of positive charge. We can use a battery and an arrangement of metal plates to set up an electric potential energy curve that has roughly the same shape as the roller coaster track. Electrons moving through this system of plates will be attracted by the positively charged plate and consequently accelerate, gaining more and more kinetic energy as they approach. Again, as for the roller coaster, this gain in kinetic energy is compensated for by a corresponding loss in potential energy, now in the form of electrical potential energy. As before, we can write down an

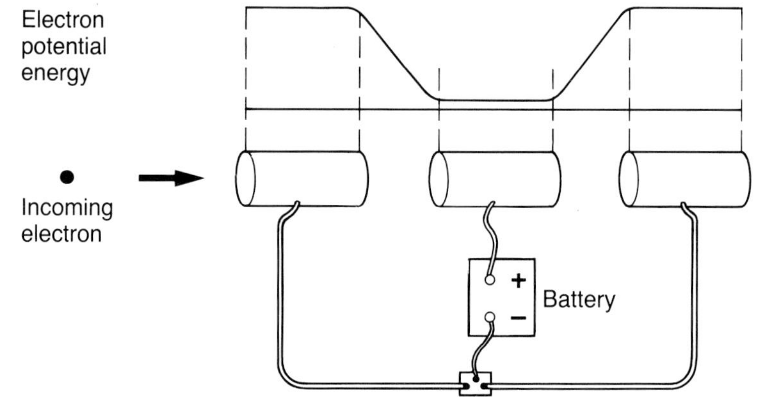

Fig. 3.4 A diagram of an apparatus to produce an electron roller coaster. A system of metal tubes is connected to a battery to give the electric potential energy curve shown at the top of the figure. An incoming electron from the left is attracted to the positively charged central tube and then overshoots as for our fairground roller coaster example.

Erwin Schrödinger was educated in Vienna, and during the First World War served as an artillery officer. After the war he decided to abandon physics and take up philosophy but the city in which he had hoped to gain a university position was no longer in Austria. Fortunately Schrödinger remained a physicist and discovered the central equation of quantum mechanics in 1926. In 1928 he succeeded Max Planck as professor in Berlin. Schrödinger left Germany after Hitler came to power and eventually became professor of theoretical physics at the Institute for Advanced Studies in Dublin, Eire.

The commemorative plaque on the wall of the original building of the Institute for Advanced Studies in Dublin, Eire. It is said that Eamon De Valera, the President of Eire, found it easier to set up a new institute for Schrödinger than to persuade his Trinity College colleagues to make him a professor. Notice that instead of 'quantum mechanics' the inscription uses the old-fashioned term 'wave mechanics'.

equation describing this book-keeping:

$$E = p^2/(2m) + V$$

where V is now the electrical potential energy.

This simple equation was Schrödinger's starting point. Using de Broglie's relation between momentum and wavelength, Schrödinger was able to guess the wave equation for a quantum object travelling in a potential. Schrödinger's equation is displayed in the box below. For an appreciation of the rest of this book it is not necessary for you to understand the mathematics of this equation in any detail. Our purpose in showing you this equation is not to frighten you. Rather, it is to convince you that there is a precise mathematical basis underlying all the hand-waving discussion of quantum phenomena that you will find in the remaining chapters of this book!

Schrödinger's equation

For the motion of a particle with total energy E, moving in one dimension, x, in a region in which there is a potential, V, Schrödinger's equation reads

$$E\psi = -\frac{\hbar^2}{2m}\frac{d^2\psi}{dx^2} + V\psi$$

It is the convention to represent probability amplitudes by the Greek letter psi (ψ). The mass of the particle is m, and $\hbar$ (pronounced 'aitch bar') is Planck's constant h divided by 2π.

Readers familiar with calculus and differentiation can see this equation solved for the problem of an electron in a simple potential in appendix 2.

Electron and neutron optics

When Schrödinger's famous paper was published in 1926, the existence of matter waves had not been experimentally established. Nowadays, the observation of wavelike behaviour for 'particles' is a commonplace occurrence and forms the basis for new ways of revealing the quantum world. Probably the most widely used device that exploits the dual particle and wave nature of matter is the electron microscope. Instead of lenses made out of glass, as in ordinary optical microscopes, an arrangement of electric and magnetic fields can be set up that act in the same way for electrons as glass lenses do for light. Why is this useful? It is useful because the amount of detail that is visible on an object that is being examined depends on the wavelength of the wave that is used to look at it. Roughly speaking, the wavelength we use must be smaller than the size of any detail that we wish to pick out or 'resolve'. The shorter the wavelength, the finer the detail that is revealed. Optical microscopes cannot resolve detail smaller than the wavelength of visible light. Thus, features smaller than about one-millionth of a metre (a micrometre or 'micron') cannot be resolved with visible light (see appendix 1). For electrons, on the other hand, the wavelength of the associated wave depends on the momentum of the electron, in accordance with de Broglie's relation. Moreover, this wavelength decreases as the momentum is increased. We are therefore able to vary the resolution of the microscope merely by varying the speed to which we accelerate the electrons. A typical electron microscope can operate at wavelengths a million times smaller than optical wavelengths. Such wavelengths are smaller than the size of atoms, and by using careful techniques some atoms can be made visible with electron microscopes. The practical resolution of electron microscopes is limited by technical problems such as defects in the

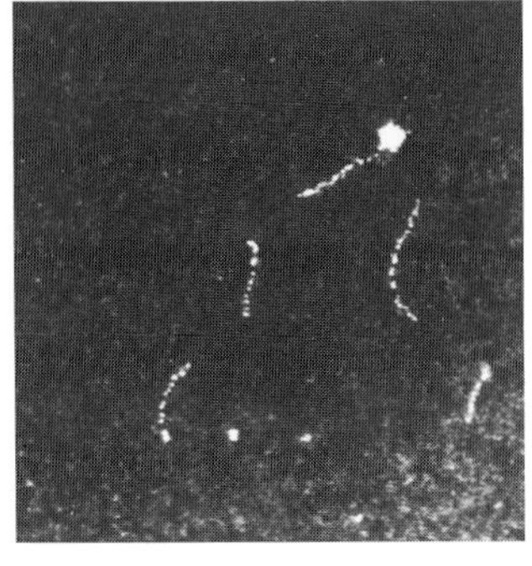

Fig. 3.5 An electron microscope picture of individual thorium atoms on a thin carbon film. The thorium atoms are visible because they contain more electrons than the carbon atoms and are able to scatter electrons more effectively. The thorium atoms are arranged in long chains by a chemical process in which each thorium atom is linked to the next by an organic molecule. In the picture, each bright spot, corresponding to a thorium atom, is separated by a distance of a little more than a thousand millionth of a metre, which is several times bigger than an individual atom.

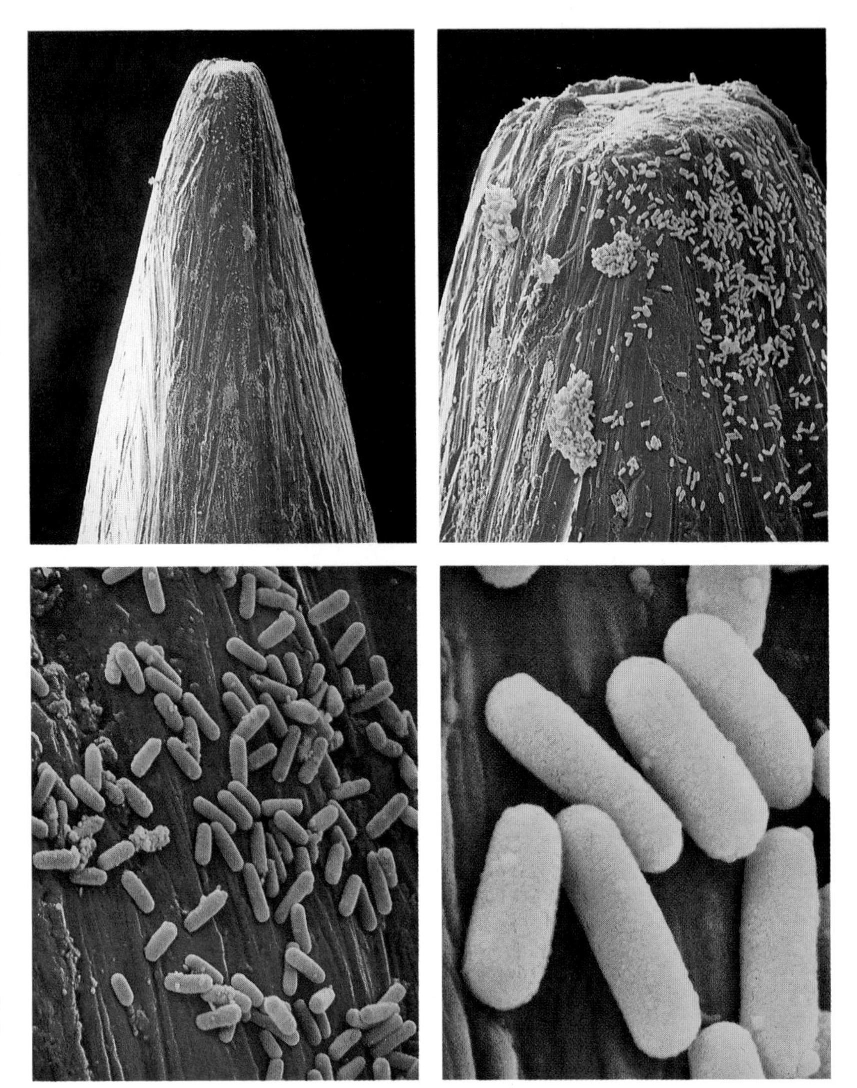

Fig. 3.6 A disturbing sequence of photographs of bacteria on the head of a pin. The magnifications are × 20, × 100, × 500 and × 2500.

Fig. 3.7 This electron microscope picture shows a family of dust mites seemingly gently grazing in a field. The magnification is about × 200.

Fig. 3.8 The SLAC two mile acclerator in Stanford, California, starts, runs from the San Andreas hills, under the freeway from San Jose to San Francisco, to the cluster of experimental laboratories at the bottom. The famous earthquake fault runs along the base of the San Andreas hills and there are elaborate safety devices to turn the accelerator off in the event of a serious earth tremor. Electrons and positrons are accelerated along the two mile length and emerge at the bottom travelling at almost the speed of light to be directed at a variety of targets to observe the collisions.

lens systems, vibrations of the apparatus and of the atoms themselves. Nevertheless, electron microscopy provides a spectacular picture of a world completely invisible to optical microscopes. Some dramatic examples of electron microscope photographs are shown in figs. 3.5–3.7.

Electron microscopes produce an image of the object being examined, and such photographs are relatively easy to interpret. However, electron beams can also be used to probe deep into the inner structure of matter without producing a direct image of the object. At Stanford in California, electrons are accelerated down a tunnel two miles long to speeds only slightly less than that of light itself. Such very high momentum electrons can explore very tiny distance scales. Observing the pattern of electrons scattered by a proton target allows us to infer details about the structure of the proton. Such electron scattering experiments, pioneered at the Stanford Linear Accelerator over the last 15 years, have enabled us to look deep 'inside' protons. The results were very surprising. Instead of finding all the positive charge of a proton being spread out uniformly throughout its volume, the experiments revealed that the charge is concentrated on even smaller constituents within the proton. Moreover, instead of these constituents having the same amount of electric charge as electrons or protons, they are now found to have charges $\frac{1}{3}$ or $\frac{2}{3}$ times this amount! These tiny constituents with their peculiar charges are called 'quarks'. They were first proposed as elementary building blocks of matter by Murray Gell-Mann, already a famous physicist at CalTech in Pasadena, California, and, at the same time, by George Zweig, then an almost unknown US physicist working at CERN in Geneva. In his paper, Zweig called these constituents 'aces', but Gell-Mann introduced the word 'quark', after a nonsense word in James Joyce's novel, *Finnegan's Wake*. Since in the quark theory of matter, the proton is made up of three quarks, the quotation from Joyce, 'Three quarks for Muster Mark!', is very appropriate and Gell-Mann's name has stuck. This is despite the fact that in German 'quark' means cream-cheese made from skimmed milk and hence is used colloquially to mean 'rubbish'! We shall meet quarks again in chapter 10 when we discuss elementary particle physics and the 'strong nuclear force'.

Murray Gell-Mann was born in 1929 and entered Yale University when he was 15. He obtained his PhD degree from MIT when he was 22 and has been at CalTech in Pasadena since 1955. Gell-Mann was awarded the Nobel Prize in 1969 for his many contributions to particle physics, not the least of which was the idea of quarks as fundamental building blocks of matter.

There is another interesting side story about quarks, or rather aces, which indicates that we physicists are not as immune from prejudice as we

Fig. 3.9 A photograph of one of the enormous electron detectors at SLAC. The electron beam enters from the left and collides with protons in a target. The scattered electron is deflected by large magnetic fields and its direction and momentum measured.

George Zweig was a student at CalTech and went to CERN in Geneva, Switzerland, after he completed his PhD. It was there that, independently of Gell-Mann, he dreamed up what is now known as the quark model of elementary particles. Zweig is now working on problems in biophysics at Los Alamos.

would like to think. At the time that Gell-Mann and Zweig suggested this theory involving, as it did, new, even more fundamental particles, another theory was in fashion which could be crudely characterized by the slogan of 'nuclear democracy'. In this rival theory, no particle was any more fundamental than any other, and so deeply were most physicists committed into thinking along these lines that to propose a constituent model involving new fundamental particles seemed like heresy! Gell-Mann, in fact, realized that there would be great opposition to the quark idea and therefore made the conscious decision to publish his paper in a European journal, where he felt the prejudice would not be so great as in the USA. Zweig, on the other hand, was in Europe, but, wisely or unwisely, wanted to have his research published in a US journal. After fighting and winning a long battle with the CERN management to be able to send his paper to a US journal, his paper was eventually rejected for publication, and Zweig was branded as a charlatan or worse by some US physicists! As a result, Zweig's paper became one of the famous 'unpublished papers' of physics: it is only now, almost 20 years later, that his paper has finally been published in a compilation of influential papers on the quark model. Today in particle physics, the idea of quarks as fundamental constituents of the proton is accepted as an 'obvious truth', while nuclear democracy is now seen to have been a bold but misguided diversion.

Wavelike properties have now been demonstrated for several other 'particles' besides electrons. In particular, over the past decade, a series of beautiful experiments have been performed with neutrons which rely on their wavelike character. Neutrons, as their name implies, are electrically neutral. They weigh about the same as protons, and together with protons they form the constituents of nuclei. Neutrons are produced in the nuclear reactions that generate power in nuclear reactors, as we shall discuss in a later chapter. All that we need to know here is that beams of neutrons can be

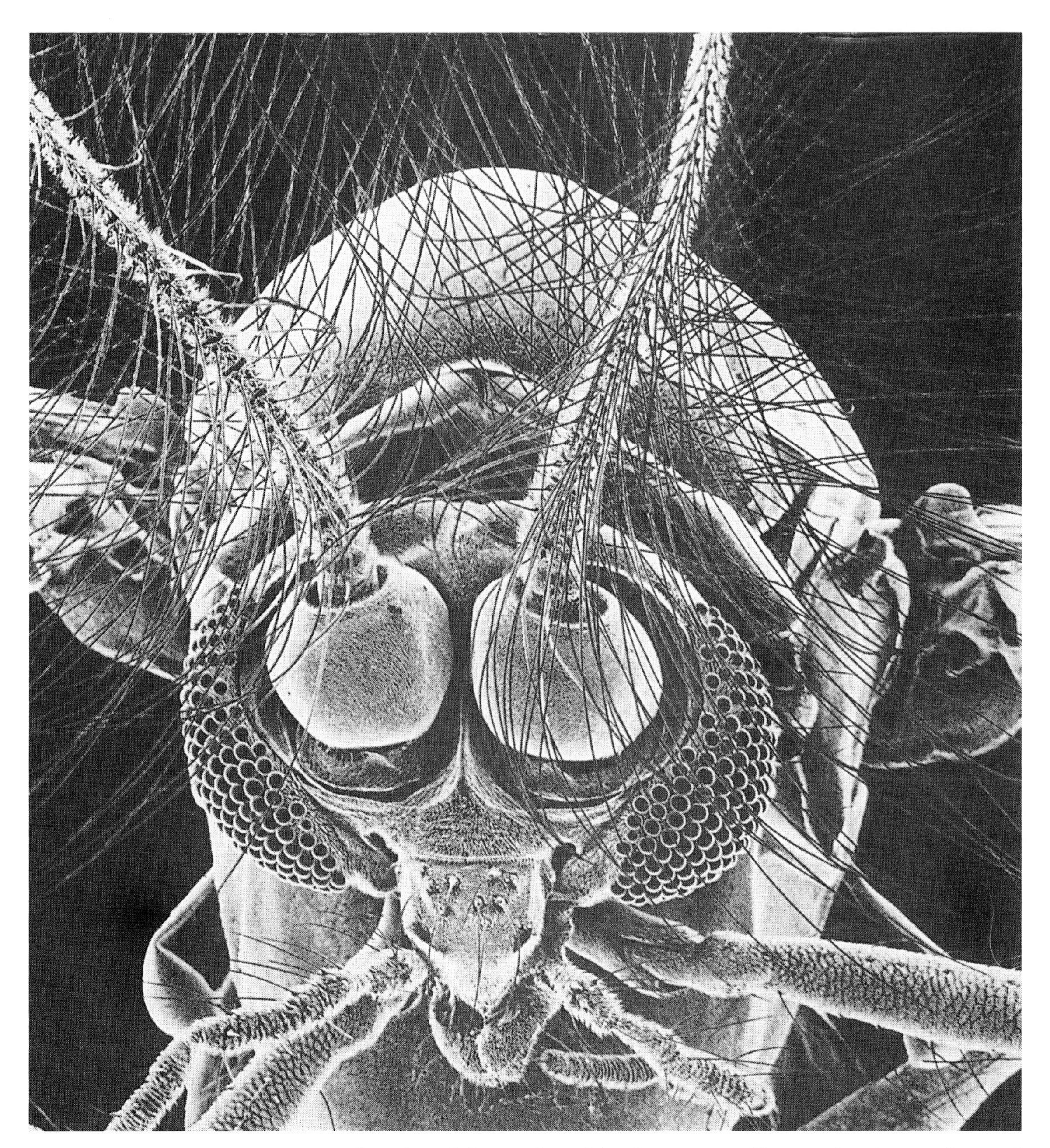

Fig. 3.10 Electron microscope picture of a common midge at a magnification of about × 500. Large swarms of these can be seen in summer but fortunately their true size is only about 2 mm.

produced, and that with such beams it is possible to perform an analogue of the double-slit experiment. For this neutron version of the experiment, instead of moving along the detector to see the interference fringes, the detector is kept in a fixed position and the interference is observed by altering the effective length of one of the two interfering paths. This is done by inserting a gas cell in one of the beams. Then, by varying the density of gas in the cell, the effective path length for neutrons can be changed, causing an interference pattern to be seen in the intensity of neutrons arriving at the detector. Neutron experiments have now become so sensitive that it is possible to observe the effects of a tiny gravitational potential term in the Schrödinger equation for neutrons.

4

Atoms and nuclei

Atoms are completely impossible from the classical point of view.

Richard Feynman

Rutherford's nuclear atom

Before quantum mechanics came along, classical physics was unable to account for either the size or the stability of atoms. Experiments initiated in 1911 by the famous New Zealand physicist, Ernest Rutherford, had shown that nearly all the mass, and all of the positive charge of an atom, are concentrated in a tiny central core which Rutherford called the 'nucleus'. Most of the atom is empty space! A table of the relative sizes of atoms, nuclei and other quantum and classical objects is given in appendix 1. Rutherford had already won a Nobel Prize earlier, in 1908, for his work on radioactivity. Radioactivity is now known to be due to the 'decay' of a nucleus of certain unstable chemical elements: some radiation is given off – in the form of alpha, beta or gamma rays – and a nucleus of a different element is left behind. As you can imagine, it took physicists some time to disentangle what was going on, and it was Rutherford who showed that the positively charged, heavy, penetrating alpha rays were, in fact, helium atoms which had lost two electrons. Beta rays, on the other hand, were identified as electrons, and gamma rays as high energy photons. At that time, any work involving the different chemical elements was regarded the province of chemists, and Rutherford was somewhat put out at winning the chemistry Nobel Prize. In his acceptance speech he remarked that he had observed many transformations in his work on radioactivity but none as rapid as his own, from physicist to chemist!

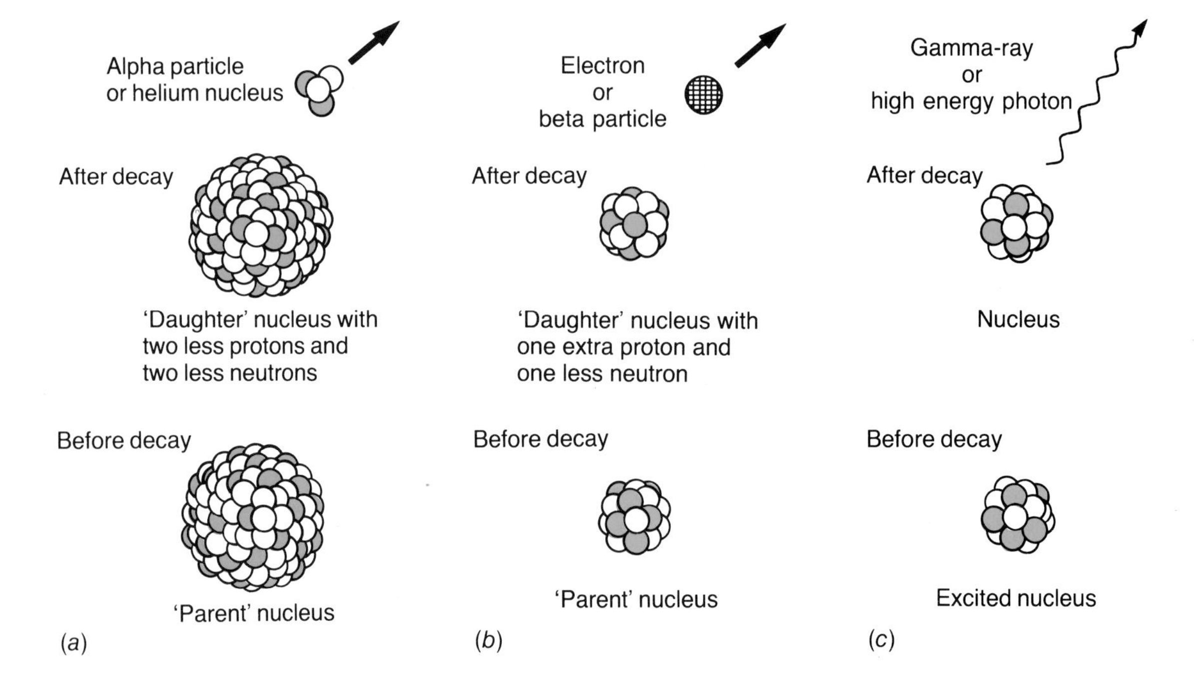

Fig. 4.1 Types of radioactivity. (a) Alpha rays are helium nuclei emitted when an unstable nucleus decays. The final 'daughter' nucleus has two fewer protons and two fewer neutrons than the original 'parent' nucleus. In this figure the black circles represent protons and the open circles neutrons. (b) A beta particle is an electron ejected from an unstable nucleus as it decays. The new nucleus has one more proton and one less neutron than the parent nucleus. (c) Gamma rays are just high energy photons that are emitted in a transition of a nucleus from an 'excited' state to a lower energy state. The number of protons and neutrons in the nucleus is unchanged.

Ernest Rutherford, first Baron Rutherford of Nelson, was born in New Zealand in 1871. He is pictured here talking to J. Ratcliffe in the Cavendish Laboratory. Rutherford had a booming voice that could upset delicate experimental instruments and the notice 'TALK SOFTLY PLEASE' was playfully aimed at him. Rutherford was one of the greatest experimental physicists of the 20th century. As well as his own fundamental research into radioactivity and nuclear physics, he influenced a whole generation of British experimental physicists.

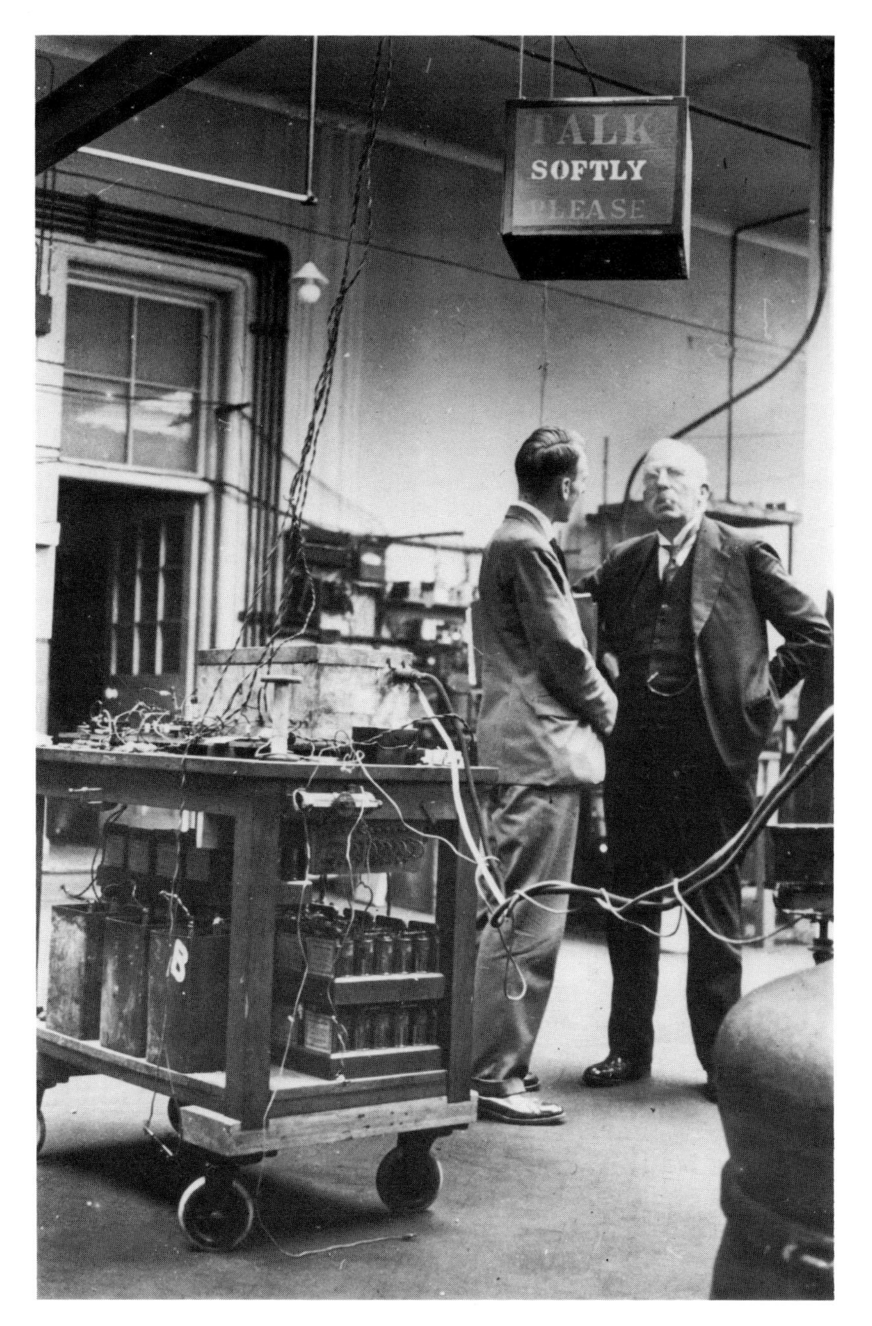

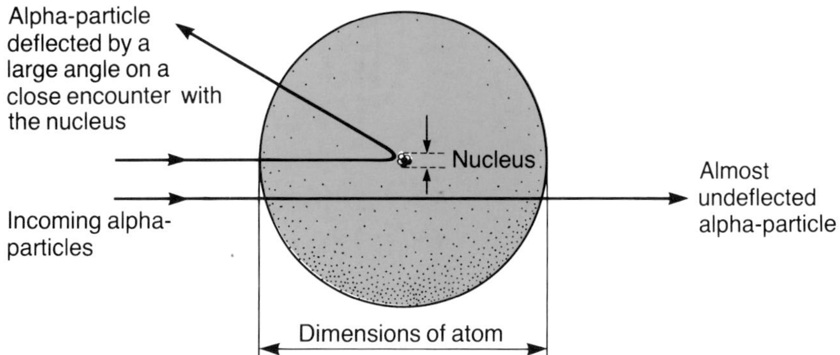

Fig. 4.2 An illustrative picture of Rutherford's alpha scattering experiment. The nucleus is about one hundred thousandth the size of an atom and its size has therefore been greatly exaggerated in this figure. Nonetheless, it is still apparent that most of an atom is empty space! Only if an alpha particle happens to collide with the tiny nucleus will there be scattering through a large angle and this will happen very rarely.

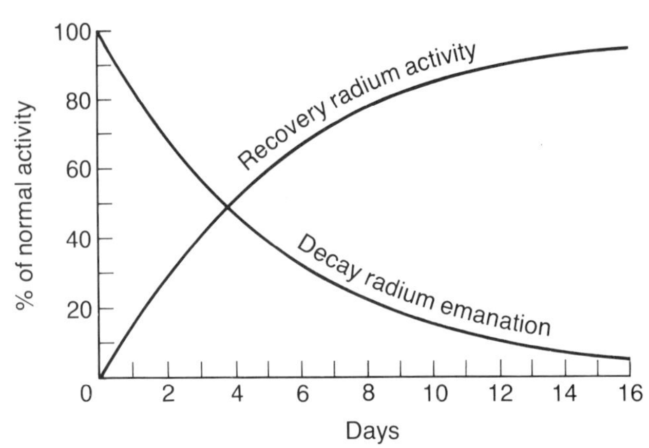

Lord Rutherford of Nelson's coat of arms. Rutherford was made a baron in 1931 and chose to link his name with the town of Nelson, in New Zealand, where he was born. The Kiwi at the top and the Maori holding a club on the right testify to his love of New Zealand. Notice the two curved lines on his shield. These were taken from a graph in a famous paper of his on radioactivity. On the left hand side of the shield is Hermes Trismegistus (thrice greatest), the name given to the Egyptian God 'Thoth', who was regarded as responsible for mysterious things such as alchemy. This is appropriate since, in a sense, Rutherford received his Nobel Prize for alchemy, albeit in a modern guise! His motto means 'To seek the beginnings of things'.

How did Rutherford discover the nucleus? He used the traditional method of physicists, namely that of throwing things at something and seeing what happens. Rutherford, together with his colleagues in Manchester, UK, fired alpha particles from a radioactive source at a very thin gold foil. They then watched carefully to see in which directions the particles were scattered. Most of the time, the alpha particles only changed direction a little, but, very occasionally, the particles were deflected through very large angles. Rutherford described his astonishment at the results in very graphic terms: 'It was quite the most incredible event that ever happened to me in my life. It was as incredible as if you fired a 15-inch shell at a piece of tissue paper and it came back and hit you!' Rutherford puzzled over these results for some weeks and eventually realized that the alpha particles could only be scattered through such large angles if they had collided with a very dense and small core of matter within the atom – the atomic nucleus.

We now know that the nucleus of an atom contains particles called protons, with a positive electrical charge, equal and opposite to the charge of the electron, and particles called neutrons that are electrically neutral. Both the proton and the neutron are about 2000 times heavier than an electron, so most of the mass of the atom resides in the nucleus. Different numbers of protons and neutrons in the nucleus then account for the different elements. The protons and neutrons are held within the tiny volume of the nucleus by forces much, much stronger than the electrical repulsion between protons. Furthermore, these 'strong forces' permit only certain numbers of neutrons and protons to combine together to form stable nuclei. The simplest nucleus is that of hydrogen as it consists of just one proton. Next simplest are the alpha particles which are helium nuclei containing two protons and two neutrons. In a neutral atom, the charge of the nucleus is exactly balanced by

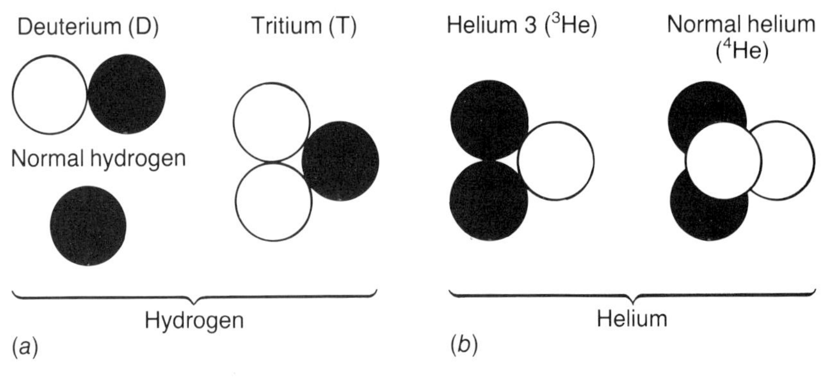

Fig. 4.3 Isotopes of hydrogen and helium. Solid circles represent protons and open circles neutrons. (a) Deuterium and tritium (b) ^{3}He and normal helium, ^{4}He.

Fig. 4.4 The emblem of the US Atomic Energy Commission showed a schematic picture of the Rutherford–Bohr model of the atom. Electrons are shown orbiting the central nucleus rather like planetary orbits in the solar system.

Niels Bohr (1885–1962) was very much influenced by Rutherford with whom he worked while he built his model of the atom. Bohr was always concerned to be very clear in his statements on quantum theory. Paradoxically this made him seem very obscure on first hearing! Nevertheless, he was undoubtedly one of the most influential scientists of this century. Bohr was regarded as the 'oracle' for questions about the interpretation of quantum mechanics and, in a celebrated debate lasting many years, he and Einstein argued about the philosophical basis for quantum mechanics. Einstein's arguments were rebutted time and time again by Bohr. Einstein remained unconvinced until the end of his life but most physicists today are content to accept Bohr's interpretation.

the charge of the electrons. Atomic hydrogen has one electron and helium two. It is the number of electrons, or equivalently the number of protons, that determines the chemical character of the different elements. Thus, although the strong nuclear force often allows a nucleus of an element to exist in several varieties, corresponding to different numbers of neutrons, all these 'isotopes' are chemically identical. For example, the gas neon is most commonly found to have a nucleus containing ten protons and ten neutrons: however, there are two other naturally occurring varieties of neon with nuclei containing eleven and twelve neutrons. Since these isotopes of neon have the same number of protons, and therefore the same number of electrons, they have the same chemical properties. Similarly, there are rare isotopes of hydrogen in which the nucleus consists of a proton and one or two neutrons. These isotopes of hydrogen are called 'deuterium' and 'tritium', respectively, and we will see later that they are important in the nuclear reactions of stars and of nuclear weapons. Some isotopes, especially those of heavy elements, are unstable and undergo radioactive 'decay' to more stable elements. We will return to these topics in a later chapter. Here, our concern is with the existence of the atom itself.

Rutherford pictured the atom rather as a miniature solar system, with electrons orbiting the nucleus, in the same way as planets circle the sun. The relatively large orbits of the electrons could then account for the large size of the atom compared with that of the nucleus. The atom as a whole is electrically neutral and the electrons are kept in orbit by the electrical attraction between them and the positively charged nucleus. Unfortunately for classical physics this whole arrangement is unworkable. To keep in orbit round the nucleus, the electrons cannot travel in a straight line – they must keep changing direction in order to keep in their orbits. In other words, they are always being accelerated towards the nucleus. But according to the well-established laws of electricity and magnetism, a charged particle which is accelerated will radiate light. Classical physics therefore predicts that in a very short time the electrons lose energy by radiation and spiral into the nucleus!

There is no answer to these problems within the framework of classical physics. More fuel was soon added to the funeral pyre of 19th century physics by a young Danish physicist called Niels Bohr. He was at Manchester with Rutherford and was bold enough to recognize that, in spite of all the difficulties, there must be some truth in the planetary model of the atom. Bohr therefore devised a sort of 'recipe book' that gave rules for calculating certain stable electron orbits for which the laws of classical physics were 'inoperative'! What gave Bohr the clue to his rules and why did physicists take them seriously? To see this, we must uncover another problem for classical physics and introduce a Swiss mathematics teacher named Johann Jakob Balmer.

Physicists had played around making electric sparks in tubes containing various gases. They found that each gas gave off light with a characteristic 'spectrum' – only certain wavelengths were present. These are called 'line spectra' and they can be used to identify different elements. In fact, the element helium was first discovered in light from the Sun: fig. 4.5 shows a view of the Sun in 'helium light'. Some line spectra are reasonably simple, such as those for hydrogen, helium and the alkali elements, but most are very complicated (fig. 4.6). Classical physics cannot even account for the stability of the atom let alone explain the details of their spectra. It is here that a rather curious character gets himself into the physics textbooks. Balmer was a mathematics teacher who, in his spare time, was obsessed with formulae for numbers. He once boasted that, given any four numbers, he could find a mathematical formula that connected them. Luckily for physics, someone gave him the wavelengths of the first four lines in the

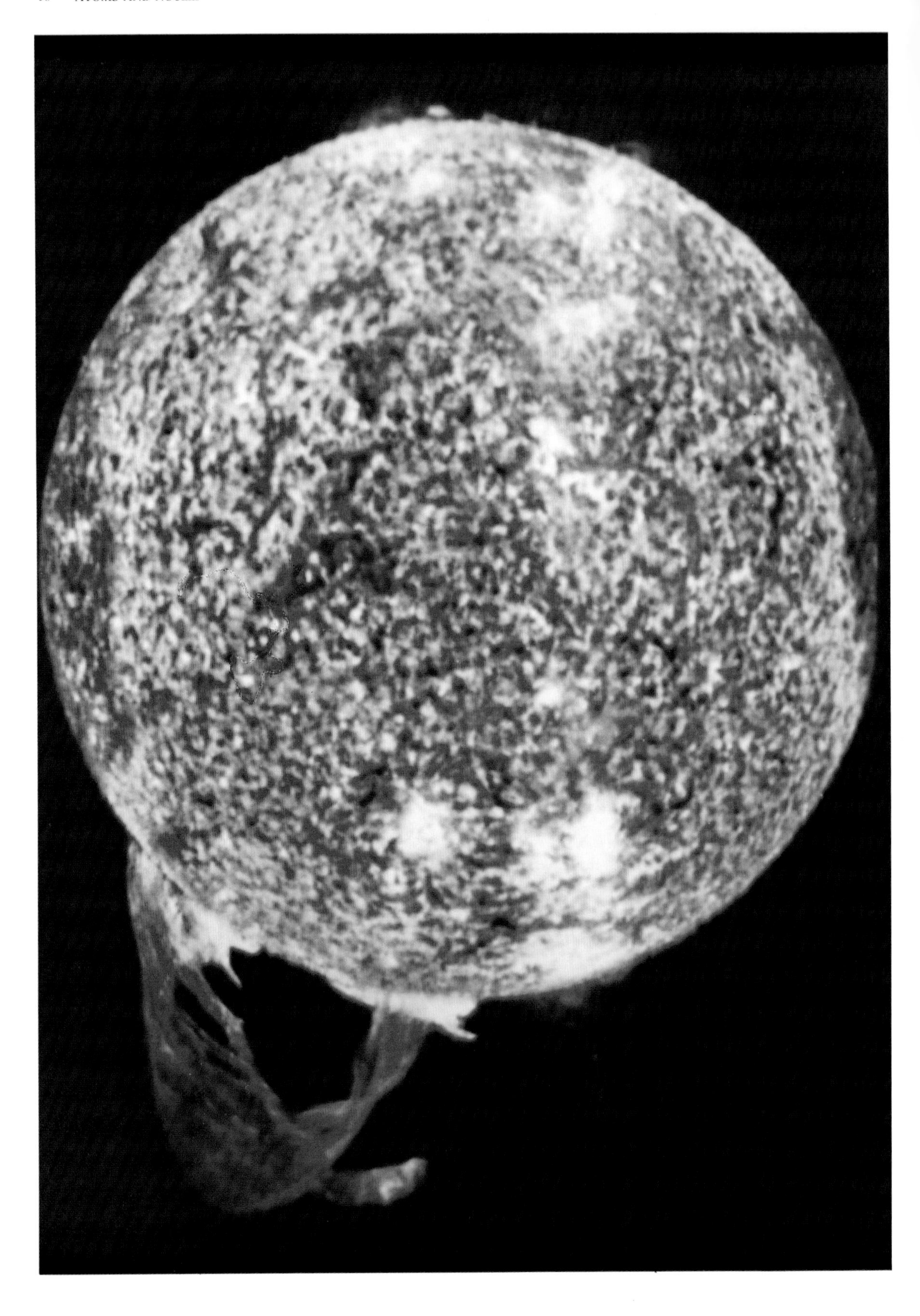

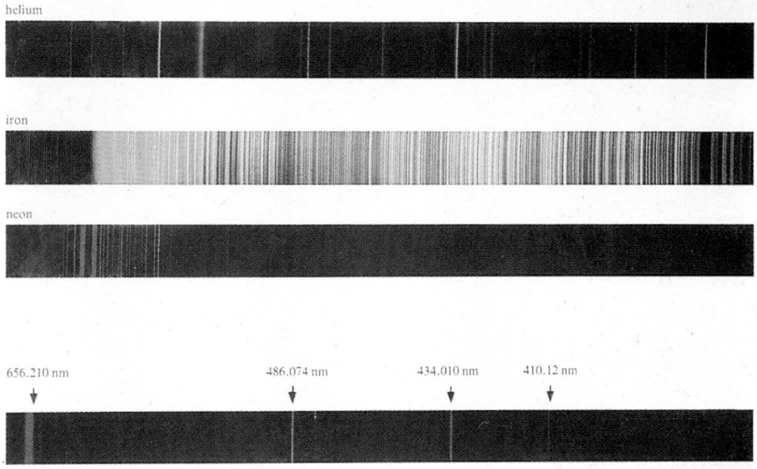

Fig. 4.6 A characteristic spectrum of light is obtained from any element in gaseous form when an electrical discharge is passed through it. This light may be separated out into the different wavelengths by passing it through a prism in a spectrometer. Each element has its own unique series of spectral lines which can serve as a kind of 'fingerprint' to establish the presence of the element. The spectrum of hydrogen is shown at the bottom. Notice how much simpler this spectrum is than that of iron.

hydrogen spectrum. The formula Balmer came up with was

$$\lambda = \frac{(364.5)\, n^2}{(n^2 - 4)}$$

where λ is in nanometres and n takes the values 3, 4, 5 and 6! The formula worked with uncanny precision but remained an unexplained curiosity until Niels Bohr was shown it. As he later said, 'Everything became clear', and Bohr's allowed electron orbits not only explained Balmer's formula but also predicted some new line spectra for hydrogen at different wavelengths. These new spectra were given by a formula of a similar form to Balmer's, except that the denominator was changed to $(n^2 - m^2)$ with m set equal to other integers besides the 2 of Balmer's original formula. After all these new spectral lines had been observed, physicists were forced to take Bohr's model seriously, despite its apparently arbitrary 'suspension' of the laws of physics. Bohr developed his model in 1913: it was not until 1926 that Schrödinger was able to explain Bohr's rules using the new quantum mechanics.

Quantized energy levels

The key to Bohr's understanding of the line spectrum of hydrogen was the idea that electrons in motion round the nucleus were only allowed in certain particular orbits. If we ignore for the moment the problem that all the orbits are unstable because the electrons should radiate away energy, then each of these orbits will correspond to a specific energy for the electron – we say that the energy is 'quantized'. This is completely at odds with our everyday experience. In our roller coaster example of the preceding chapter, we can set the carriage rocking back and forth across the valley starting at any

Fig. 4.5 The Sun emits light across the whole electromagnetic spectrum. Different parts of the spectrum tell us about different aspects of the processes occurring in the Sun. It is particularly revealing to use a filter that only passes light with a wavelength corresponding to a spectral line of one particular element. In this picture an ultra-violet line from helium has been used to construct a picture of the Sun in helium light. The picture has been coloured by computer with the yellow parts representing the regions where there is most intense emission. This picture was taken by Skylab and picks out a region in the lower atmosphere with a temperature between 10 000 and 20 000°C. As well as showing the Sun as rather blotchy, the picture also shows a spectacular arch of material propelled away from the Sun by magnetic forces.

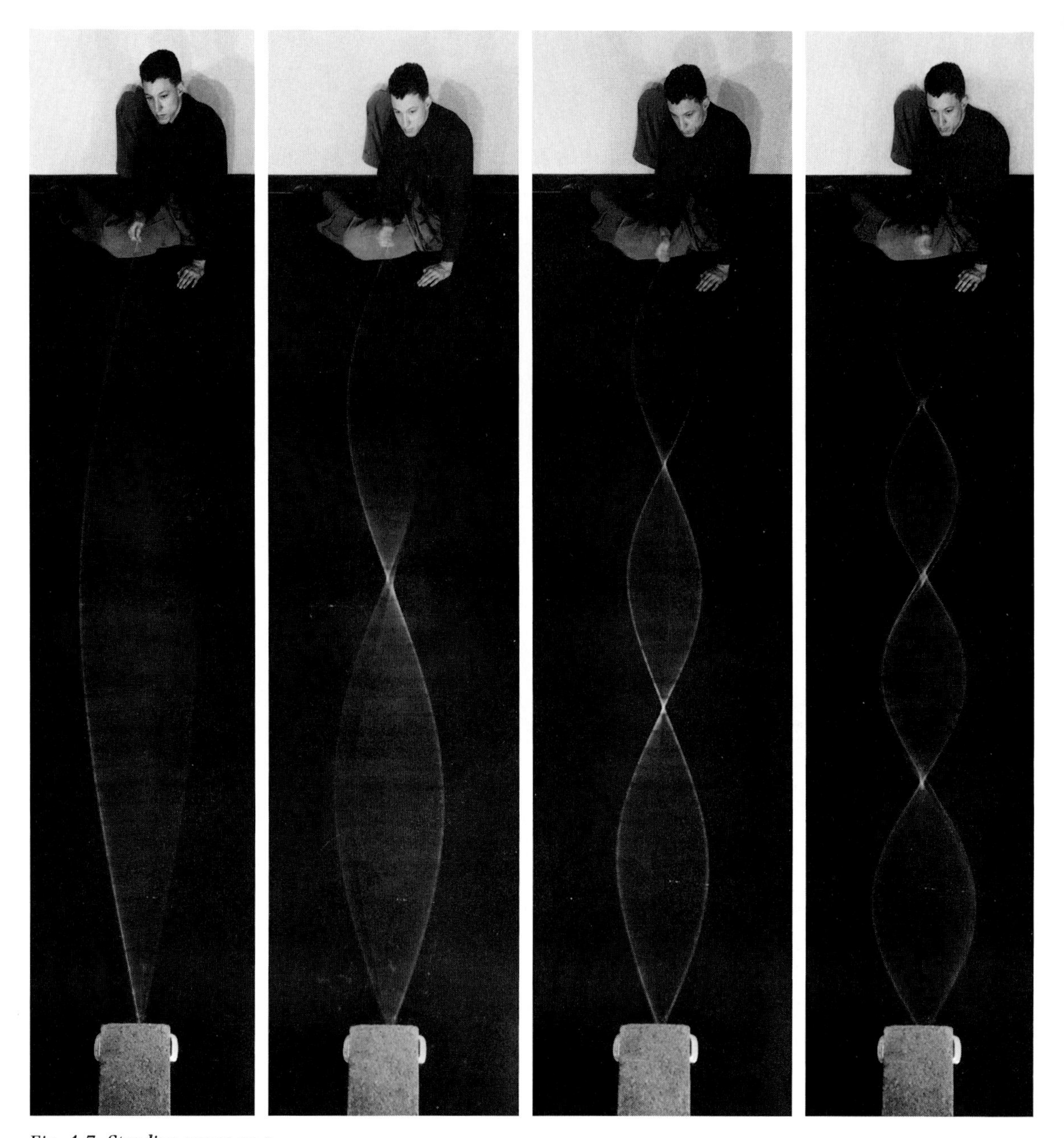

Fig. 4.7 Standing waves on a string. The photographic exposure is longer than the period of vibration so that the photograph records most strongly the positions where the string is moving slowest. Notice that there are some positions where the string is always at rest.

height we want – equivalent to starting with any value of the total energy (fig. 3.3). How then does quantum mechanics lead to energy quantization and stable orbits?

The answer to both these problems lies with the wavelike properties of the electrons. According to quantum mechanics, the allowed energies for the electron are found by solving the Schrödinger wave equation with the appropriate potential energy term. Fortunately we can see how energy quantization comes about without going to the trouble of solving the Schrödinger equation. Imagine a potential like our roller coaster example but where the hills are very high and steep, and the valley is wide and flat, so that we have a sort of 'box' for the electron to sit in. The problem of finding the allowed energies of this quantum system is now identical to the classical

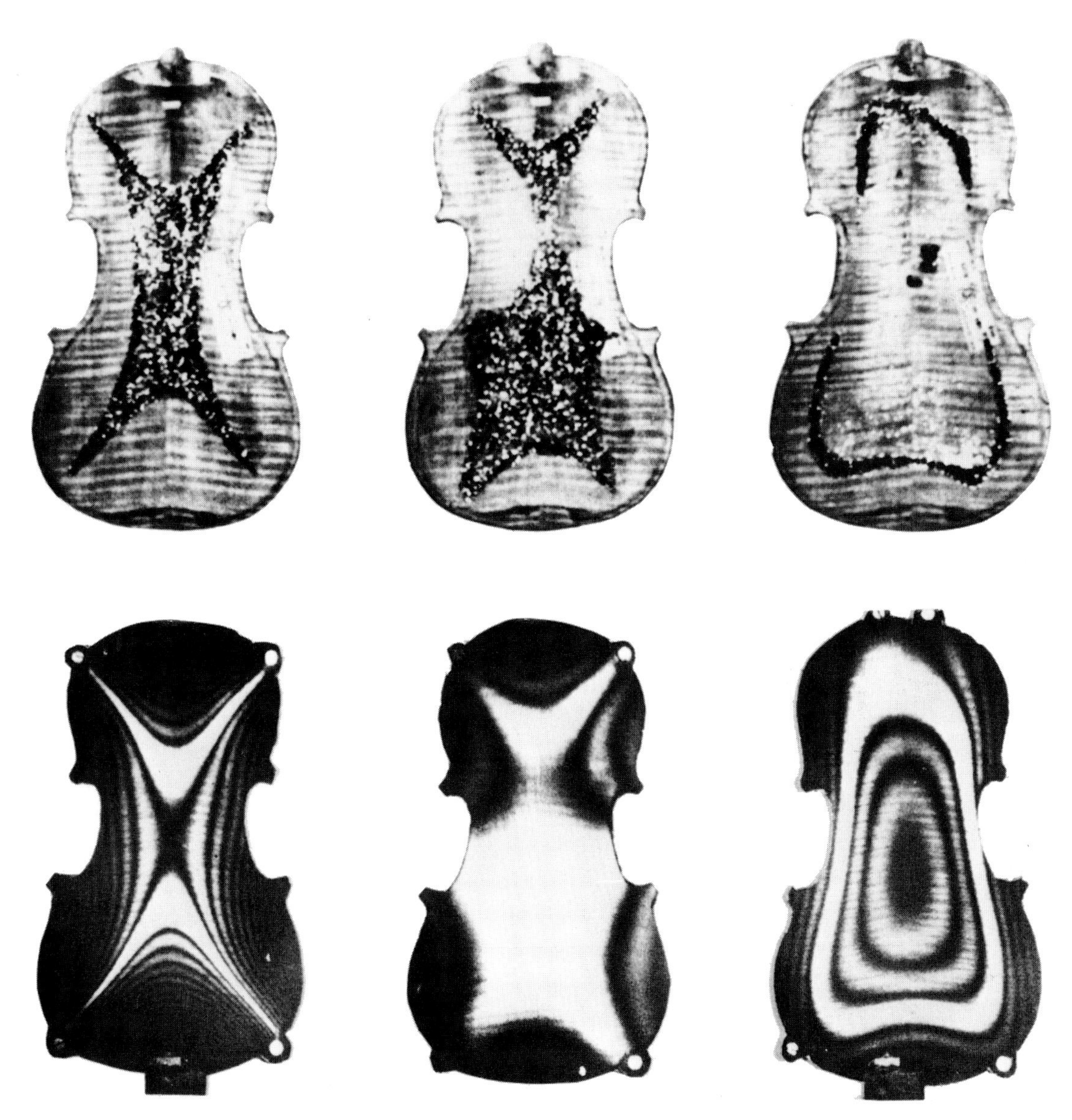

Fig. 4.8 Vibration patterns for a violin. The top three pictures are photographs of the patterns made by a sprinkling of light powder which accumulates on and around regions of little or no vibration. Below these, the same patterns are revealed by a laser interference method. The white areas correspond to regions of little vibration. The interference method is much the more sensitive.

problem of finding solutions of the wave equation for waves on a string that is fixed at both ends. For the string, it is obvious that only certain 'wavelengths' can 'fit in' between the fixed end points. Furthermore, in the case of the string, it is these wavelengths – the 'fundamental', or longest wavelength, together with the higher 'harmonics', or shorter wavelengths – that determine the sound that we hear. In quantum mechanics, it is the electron probability amplitudes that are required to fit in the box, but now each allowed wavelength corresponds to a definite electron energy. This is the origin of Bohr's quantized energies. Classically, a ball in a box can be in motion with any energy: quantum mechanically, an electron in a box can only have certain allowed values of energy.

This example of an electron in a box illustrates several general features of

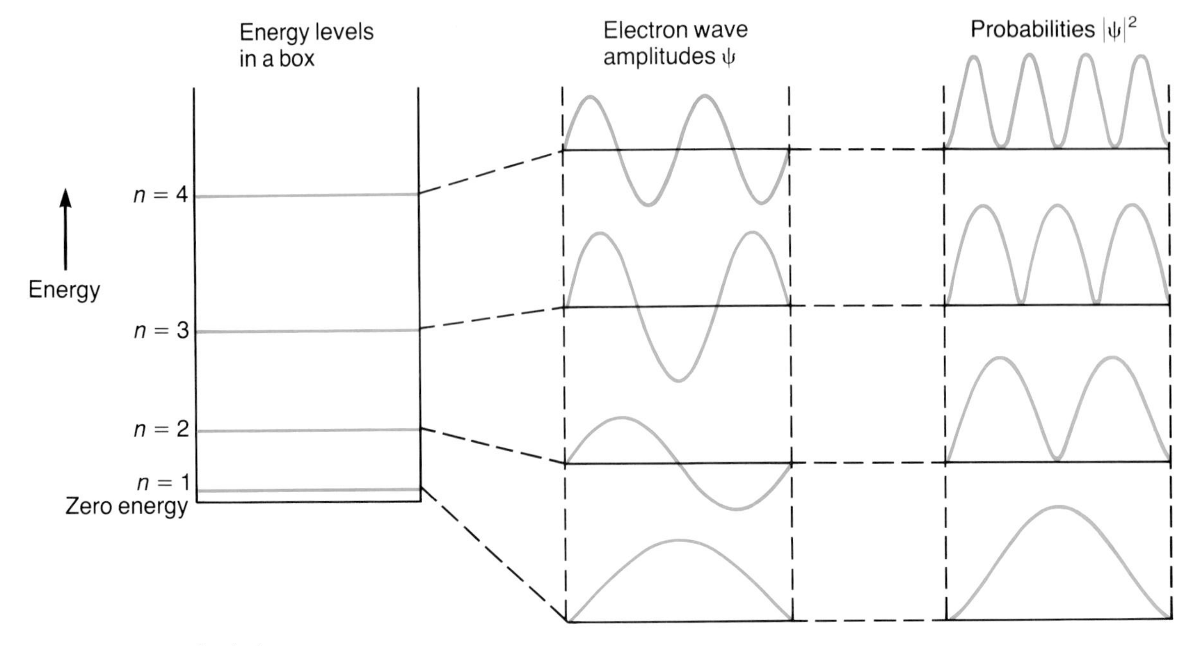

Fig. 4.9 Energy levels for a quantum particle in a box. The middle picture shows the corresponding wave patterns and the picture on the right the probability patterns for the particle. These are just the squares of the wave patterns.

quantum mechanics. The mathematical details are contained in appendix 2, but the forms for the electron probability amplitudes can be guessed from our string analogy. In fig. 4.9 we show these 'wavefunctions' together with a scale showing the corresponding electron energy. The first thing to notice is that the lowest energy of the electron is not zero. Since the uncertainty in position of the electron cannot be greater than the size of the box, Heisenberg's uncertainty principle requires a certain minimum energy. Even when the electron has its lowest possible energy – we say the electron is in the 'ground state' – it cannot sit still but must be forever jiggling around! This so-called 'zero-point' motion is a general property of quantum systems: it explains why liquid helium does not freeze solid even at temperatures close to absolute zero. The next thing to notice is that the probability amplitude for the electron in the box vanishes not only at the ends, but also, for the higher energy states, at places in between. In the case of a vibrating string these 'nodes' – places where there is no movement of the string – cause no surprise. For the electron, however, these nodes are places where there is no chance of finding the electron! The relative probability of finding the electron at different positions within the box is just given by the square of the quantum probability amplitude (see fig. 4.9). We therefore see that not only are the allowed energies 'quantized', but also the probability of finding the electron varies with position within the box and is different for different electron energies. All this is very different from our everyday intuition about particles, but follows when we admit electrons to have wavelike properties.

There is one last moral to take from this example. In order to talk about quantum energy levels and their corresponding wavefunctions it is convenient to label them in some way. Thus, we assign a 'quantum number', $n = 1$, to the ground state, $n = 2$ to the first excited state, and so on. In this example, this labelling seems almost a triviality, but the use of quantum numbers to label energies and quantum amplitudes is a general feature of quantum mechanics. Most problems of interest for applications in the real world are not, of course, as simple as this example of an electron in a box, and finding the energies and wavefunctions is often very difficult. Nonetheless, the same general principles hold true, and solving the Schrödinger equation for realistic situations is similar to the problem of

Fig. 4.10 A kettledrum sprinkled with powder reveals six of its many vibrational patterns. The powder collects near the 'nodes' where the vibration is weakest. These patterns are analagous to the quantum probability patterns for electrons in a box.

finding the vibrational wave patterns for more complicated objects than strings. Some wave patterns for some familiar objects are shown in figs. 4.8 and 4.10. There is one more lesson that we must learn from these more complicated vibrations that does not show up in our box example. In fig. 4.11 we show the vibrational modes of a square drum: the quantum problem of an electron in a two-dimensional box has just the same solutions. If we try to label the wavefunctions according to their energy we now find a problem. For the lowest energy state there is just one possible wave-function, which we can again label by a single quantum number $n = 1$. For the first 'excited state', however, we have a choice. If we label the two directions x and y, we see that we can either excite the x motion to its first harmonic and leave the y motion in the fundamental, or vice versa. Both these possibilities have exactly the same energy. We therefore need another quantum number to distinguish the two possible wavefunctions corre-

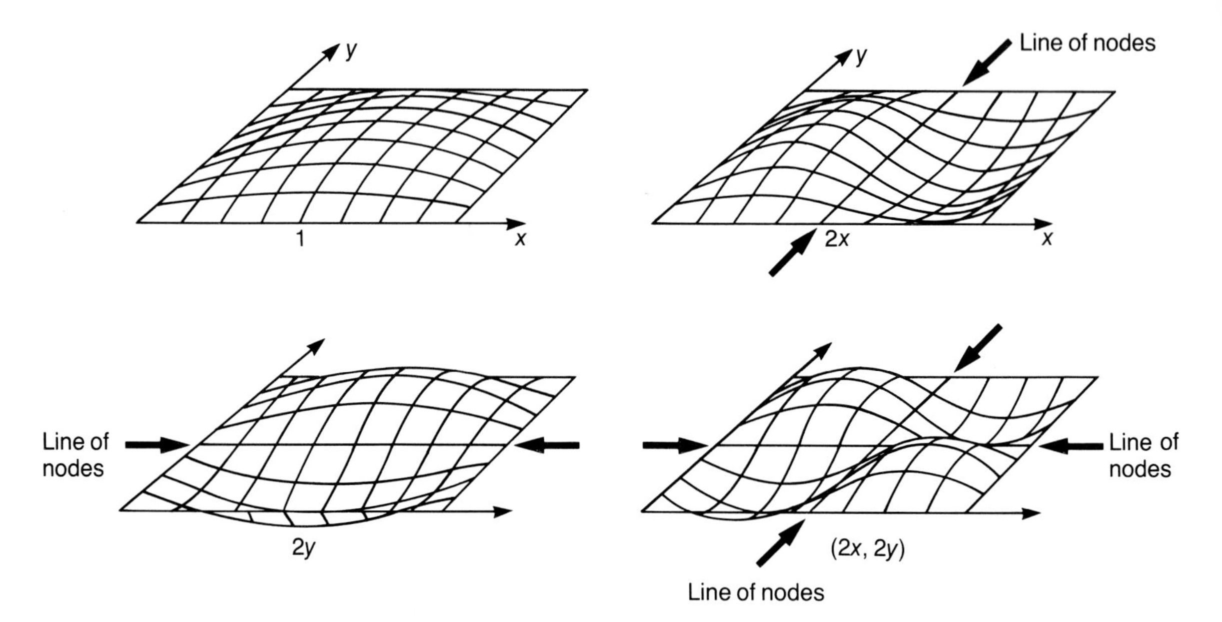

Fig. 4.11 A diagram showing the vibration patterns of a square drum. The surface of the drum has been indicated by a mesh of strings to try to make clear the correspondence with the modes of the vibrating string.

sponding to this energy. In the case of the square box, we can label them both $n = 2$ and add a label x or y to specify which direction is 'excited'. Thus, we can label the wavefunctions as $2x$ and $2y$. Physicists say that this situation – in which there is more than one possible quantum state with the same energy – is 'degenerate'. This is yet another example of physicists using an everyday word in a special technical sense! We shall find similar degeneracies when we talk about the hydrogen atom wavefunctions. In three dimensions, it should come as no surprise that we shall need at least three quantum numbers to label all the quantum states.

The hydrogen atom

One of the reasons for the almost immediate and universal acceptance of Schrödinger's equation was that, after decades of fumbling around in the dark, physicists were at once more able to calculate using standard mathematical techniques. Instead of having to follow Bohr's mysterious rules, the energy levels of hydrogen appeared naturally as the allowed frequencies of a wave problem in three dimensions. Indeed, what was astonishing was the remarkable accuracy of these predictions. When giving lectures on the new quantum mechanics, the famous Italian physicist Enrico Fermi used to say 'It has no business to fit so well!' Nevertheless it does, and we are now in a position to understand both Bohr's quantum rules and the stability of atoms.

The hydrogen atom is a quantum system consisting of a relatively massive proton, with positive charge, accompanied by a very light, negatively charged electron. The electron is attracted to the proton by a force that becomes stronger the closer the electron is to the proton. The appropriate potential energy for this situation depends only on the separation of the electron and proton and is the same in all directions. Classically, there would be nothing to stop the electron lowering its energy as much as possible until it was sitting right on top of the proton. In quantum mechanics we know that Heisenberg's uncertainty principle prevents this. The energy levels are found by solving the Schrödinger equation with this potential, and although the mathematics is much more complicated, the

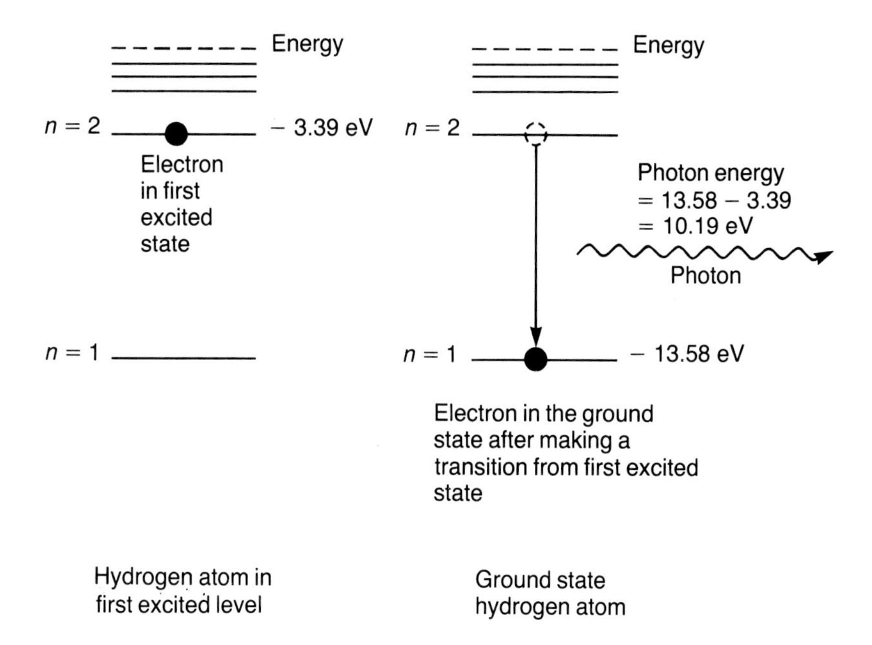

Fig. 4.12 Electron transition from the first excited state of hydrogen to the ground state with emission of a photon.

resulting energy level spectrum is similar to that of the electron in a box. With the aid of Planck's famous formula relating the energy of a photon to its frequency

$$E_{\text{photon}} = hf$$

that we talked about in chapter 2, we can now understand Balmer's magic formula. Left to itself, the electron in the hydrogen atom likes to have the lowest possible energy, and the electron therefore occupies the 'ground state' corresponding to the energy quantum number $n = 1$. However, if the atom is disturbed – by collisions with other atoms or by light shining on it – the electron may be 'excited' to a higher energy level with a larger n value. Since this state of the atom has more energy than usual, after some time the atom will 'decay' back to the ground state. Picturesquely we can say that the electron 'jumps' down to a lower energy level. To conserve energy, the excess energy is given off in the form of a light photon whose energy is given by the formula

$$E_{\text{photon}} = E_{\text{initial}} - E_{\text{final}}$$

Since frequency and wavelength of light are related by the classical wave formula

$$c = f\lambda$$

velocity of light = frequency times wavelength

we now have a prediction for the wavelengths of spectral lines. Schrödinger's result was exactly the same as that of Balmer and Bohr, namely

$$1/\lambda = \mathrm{R}\left(\frac{1}{n_{\mathrm{f}}^2} - \frac{1}{n_{\mathrm{i}}^2}\right)$$

where R is a calculable combination of constants like the mass and charge of the electron, Planck's constant and so on, and n_{f} and n_{i} are the energy quantum numbers of the final and initial levels, respectively. Fig. 4.13 shows how the various spectra come about.

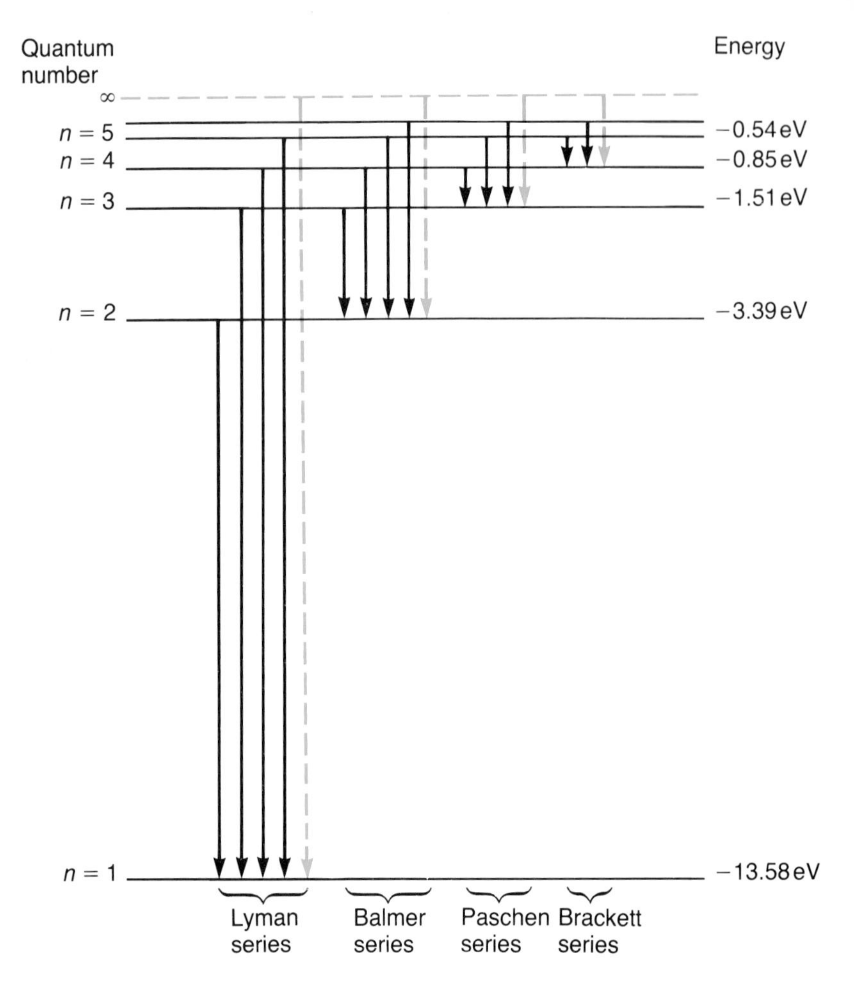

Fig. 4.13 Energy level diagram for hydrogen showing how the various 'series' of spectral lines arise. Only the Balmer series is in the visible range of wavelengths.

The same picture of the energy levels explains how light is absorbed by atoms. For light to be absorbed, not only does the light photon have to have an energy equal to some energy level difference, but also an electron must be in the right state to absorb a photon with this energy. At normal temperatures the energy of collisions between atoms in a gas is usually insufficient to excite many atoms, since there is a large energy difference between the ground state and the first excited state. Hence, at room temperatures, most atoms are in the ground state. Now, the energy differences from the ground state to any excited state are so large that the associated photons are in the 'ultra-violet' range of frequencies (see appendix 1) rather than the visible. Thus, visible light can pass through many gases unabsorbed, since almost all the atoms require a much larger photon energy for excitation from the ground state. This is the reason why most gases are transparent to visible light.

We can also explain the beautiful colours of nebulae observed in astronomical photographs. Fig. 4.14 shows a picture of a large gas cloud in our Galaxy called the Orion nebula. Inside the nebula hot stars are constantly giving out large numbers of ultra-violet photons. These photons are so energetic that they can knock the electron right out of the hydrogen atom leaving a positively charged 'ion'. As the electrons and protons recombine, the electron loses energy by cascading down through the energy levels giving off photons on the way. The red colour of the cloud corresponds to a particular electron jump – or, in more technical terms, a 'photon transition' – that appears in the hydrogen spectrum.

Fig. 4.14 Photograph of the Orion nebula, a large glowing cloud of hydrogen, inside which are many newly formed stars.

Wavefunctions and quantum numbers

We have, so far, only discussed the quantum numbers that arise in very simple situations. For a detailed understanding of the quantum mechanics of the elements that we shall discuss in chapter 6, we need to know more about the wavefunctions and quantum numbers that are needed for the hydrogen atom. Since this involves a rather detailed discussion it may be best to skim this section quickly and not get bogged down! With this warning let us now look at the electron probability amplitudes – hydrogen wavefunctions – that correspond to the energy levels of the hydrogen atom. The wavefunction of the ground state, with energy quantum number $n = 1$, turns out to be very smooth and symmetrical, and looks the same from all directions. The probability of finding an electron is proportional to the square of this wavefunction, and fig. 4.15 gives an idea of how this probability varies over a slice through the centre of the atom. Moreover, these pictures of 'probability density' do not change with time. This solves the problem of why the electron does not radiate. The electron is not described as a particle whirling round the proton but rather as a stationary pattern of probability in which nothing is being accelerated. If we look at the excited energy levels in more detail, we see, as in the square drum example, that there is degeneracy – several wavefunctions correspond to the same energy. We therefore expect that we will need several quantum numbers to label each of these wavefunctions. These 'extra' quantum numbers correspond to the quantization of another classical quantity – 'angular momentum'. In the next few paragraphs we will try to describe these new quantum numbers in a bit more detail: if this proves too heavy going, skip to the next chapter!

Angular momentum is, as its name suggests, related to ordinary momentum, and it is important in problems involving particles in orbit around some centre, such as the motion of the planets round the Sun. Imagine tying a ball to a piece of string and whirling it round and round holding the other end of the string. The angular momentum of the ball is just the ordinary momentum times the length of the string:

$$L = rp$$

angular momentum = length of string times ordinary momentum

For a fixed length of string, the faster the ball goes the more angular momentum the ball has. Angular momentum is important because, like energy, it is conserved in both classical and quantum systems. In the case of our ball on a string, imagine shortening the string as the ball whirls around. Angular momentum is conserved, and therefore as the length, r, of the string decreases, the ordinary momentum, p, of the ball must increase and the ball goes round faster. In the quantum mechanical treatment of the hydrogen atom angular momentum is conserved just as in this classical example. However, in the quantum case we are not permitted to have any value we want for the angular momentum. Quantum mechanical angular momentum is quantized, like the energy. In fact, this was how Bohr guessed his stable orbits: the angular momentum was only allowed to take values which were whole units of Planck's constant h divided by 2π. Curiously, although Bohr had the correct energy levels, Schrödinger's solution of the hydrogen atom showed that Bohr's guess about angular momentum was not quite right. Nevertheless, the angular momentum is indeed quantized and is described by two new quantum numbers, L and M. Thus, at the $n = 2$ energy level there are four degenerate wavefunctions which give rise to the probability 'slices' shown in fig. 4.16. The distribution labelled '$n = 2$, $L = 0$, $M = 0$' has angular momentum zero. This slice also has the same circular symmetry as the ground state wavefunction with quantum numbers '$n = 1$, $L = 0$, $M = 0$', and also has zero angular momentum. The other three wavefunctions all have one unit of angular momentum and angular momentum quantum number '$L = 1$'. For $L = 1$ there are three possible values of the second angular momentum quantum number M, namely $M = +1, 0$ and -1. These three possibilities correspond roughly to the three possible directions for the axes of rotation of the electron. Imagine the ball on a string again and picture it rotating about the vertical 'z axis' marked on the photographs. The ball will therefore be swinging round in a plane containing the x and y axes. Quantum mechanically this situation corresponds to the $M = +1$ state for which the probability 'lobes' are mainly in the x–y plane. The $M = -1$ state corresponds to the axis of rotation being along the negative z axis and again the probability lobes will be in the x–y plane. The $M = 0$ state with the probability 'lobes' pointing along the z direction, corresponds roughly to having an axis of rotation somewhere in

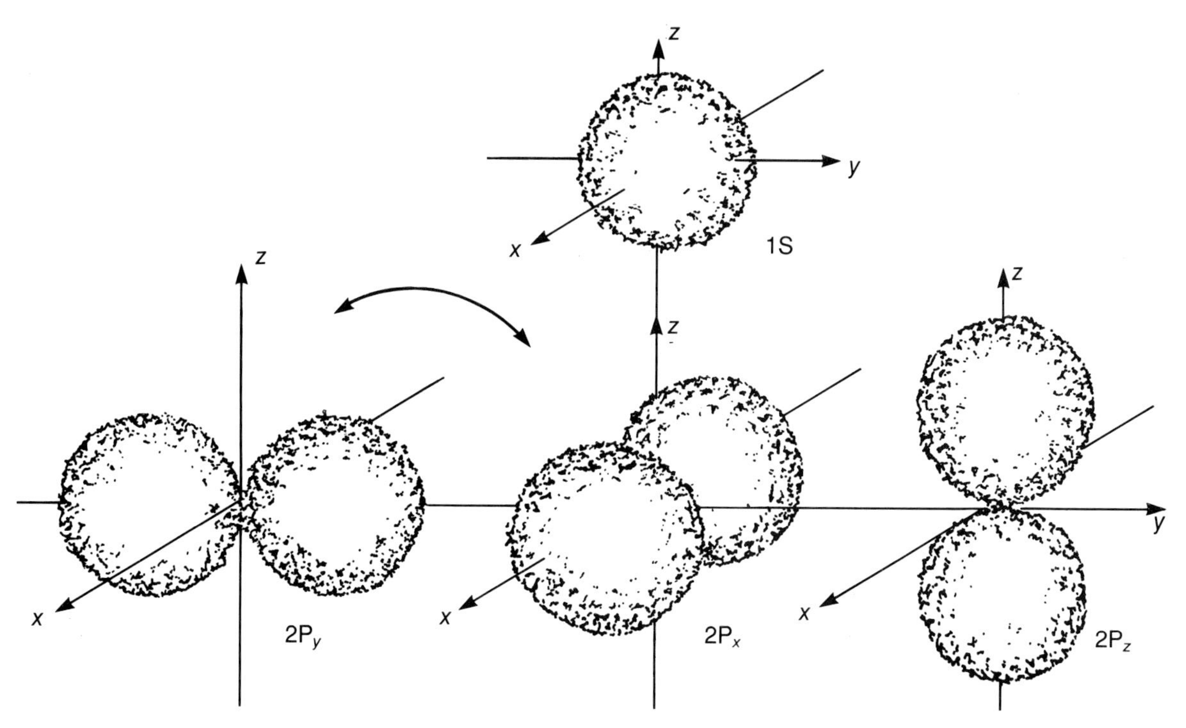

Fig. 4.15 Three-dimensional visualization of equal probability surfaces for the lowest-lying hydrogen energy levels. The 1S state is spherically symmetric and has zero angular momentum. The three 2P states all have one unit of quantum angular momentum and have probability surfaces as shown. The two 'lobes' can point in any of the x, y *or* z *directions, corresponding to the* P_x, P_y *and* P_z *states.*

the x–y plane. For understanding the chemistry of molecules, it often turns out to be more convenient to use a slightly different choice of labelling for these three $L=1$ states. If we take combinations of these M quantum number wavefunctions, we can arrange to have three degenerate wavefunctions whose 'probability lobes' actually point along the x, y and z directions. Instead of the quantum number M, therefore, we shall often refer to the three $L=1$ states by the labels x, y or z.

At the next energy level, states with two units of angular momentum, $L=2$, are present, along with $L=1$ and $L=0$ states. There are five possible M values for the $L=2$ wavefunctions: $M=+2$, $+1$, 0, -1 and -2, corresponding roughly to the axis of rotation going from the positive z direction round to the negative z direction.

It is remarkable that an understanding of these three quantum numbers provides almost all we need for an understanding of Mendeleev's periodic table of the elements (discussed in chapter 6). However, there is one respect in which hydrogen, with its single electron, is rather a special case. For atoms with more than one electron, the energies of states with the same value of n but different values of L are not the same: the energy depends on both n and L.

A last word on notation. The alkali metals lithium, sodium and potassium have line spectra that are similar to those of hydrogen. Because the physicists who first studied these spectra had no idea as to their origin, they labelled the different series they observed in a fairly arbitrary way: S for sharp, P for principal, D for diffuse and F for fundamental. We now know that these series of lines result from transitions to final states with different values of angular momentum: the sharp to $L=0$, the principal to $L=1$, the diffuse to $L=2$ and the fundamental to $L=3$. Thus, physicists usually refer to angular momentum states not by the actual L values, $L=0$, $L=1$, $L=2$ and $L=3$, but instead by this incredibly obscure historical labelling scheme, S, P, D and F! For example, it is common for the three $L=1$ states to be labelled P_x, P_y and P_z instead of by the much more obvious L and M notation.

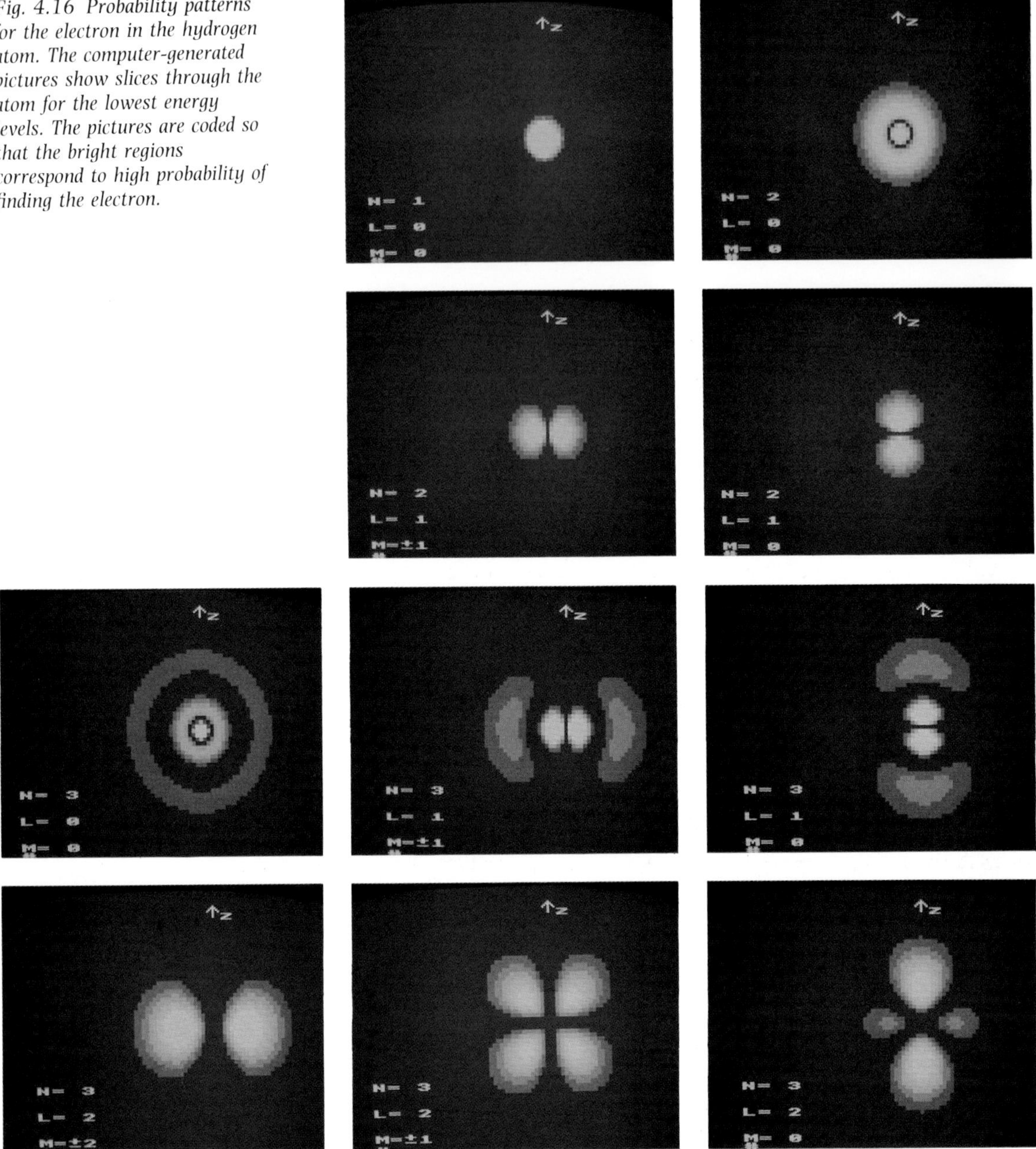

Fig. 4.16 Probability patterns for the electron in the hydrogen atom. The computer-generated pictures show slices through the atom for the lowest energy levels. The pictures are coded so that the bright regions correspond to high probability of finding the electron.

5

Quantum tunnelling

It is possible in quantum mechanics to sneak quickly across a region which is illegal energetically.

Richard Feynman

Barrier penetration

One of the most startling consequences of de Broglie's wave hypothesis and Schrödinger's equation was the discovery that quantum objects can 'tunnel' through potential energy barriers that classical particles are forbidden to penetrate. To gain some idea of what we mean by an energy barrier, let us go back to our roller coaster and look at a larger section of track, as shown in fig. 5.1. If we start the carriage from rest, high up on the left, at A, and ignore any small frictional energy losses, we know from the conservation of energy that we shall arrive on the other side at the same height we started from, at C. As we went over the little hill B, at the bottom of the valley, the car slowed down as some of our kinetic energy was changed to potential energy in climbing the hill, but because we started much higher up, we had plenty of energy to spare to get us over the top. However, if we started the carriage from rest at A, we do not have enough energy to climb over the hill D and get to E. This is an example of an 'energy barrier', and we can say that the region from C to E is 'classically forbidden'.

What is remarkable about quantum 'particles' is that they do not behave like these classical objects. An electron travelling on an 'electron roller coaster' of the same form as the roller coaster of fig. 5.1 can 'tunnel through' the forbidden region and appear on the other side! This 'barrier penetration' or 'quantum tunnelling' is now a commonplace quantum phenomenon. It forms the basis for a number of modern electronic devices such as the tunnel diode and the Josephson junction, of which more later. How can we obtain some understanding of how such tunnelling comes about, without presenting a detailed solution of the Schrödinger equation? One way of thinking about it uses an argument based on Heisenberg's uncertainty principle. In chapter 2 we phrased this in terms of the uncertainties in position and momentum measurements. However, another equivalent relation exists between uncertainties in measurements of time and energy

$$(\Delta E)\ (\Delta t) \approx h$$

Thus, although classically we can never change the total amount of energy

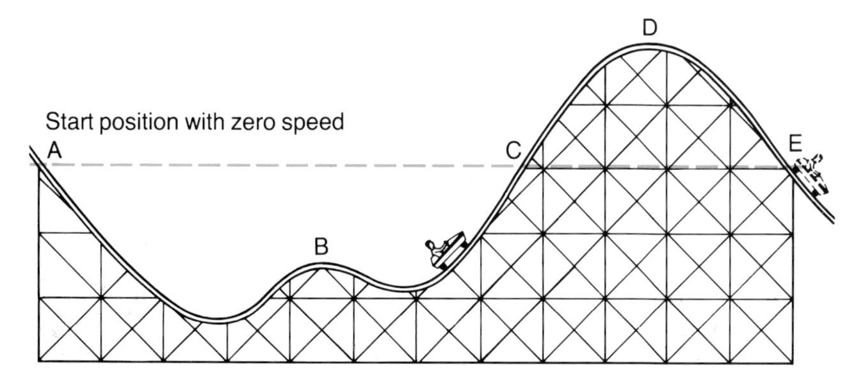

Fig. 5.1 An illustration of what quantum tunnelling means for a real roller coaster. If the carriage starts from rest at position A, conservation of energy does not allow it to go higher than position C on the other side of the valley. In quantum theory, on the other hand, there is a chance that the carriage could 'tunnel through' the forbidden region between C and E, and emerge on the other side of the hill. For a real roller coaster this tunnelling is extremely unlikely!

Fig. 5.2 In Mr Tompkins' wonderland, where Planck's constant is much larger, his car could tunnel through the wall, 'just like a good old ghost of the middle ages'.

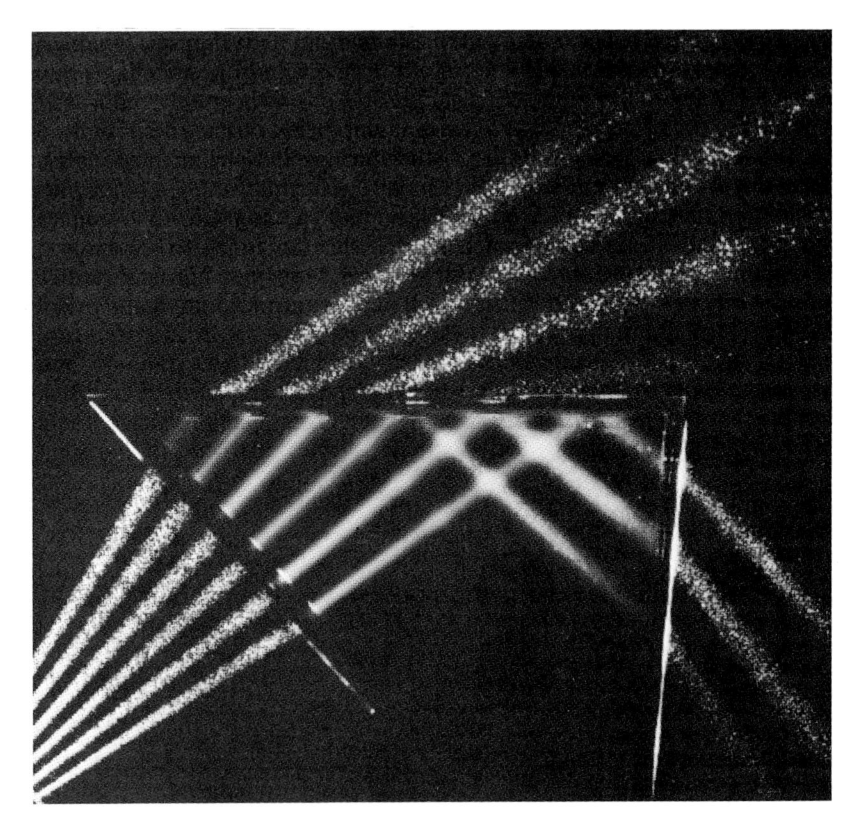

Fig. 5.3 This photograph shows several rays of light striking a prism at a variety of angles. As can be seen, beyond a certain 'critical' angle the light rays are entirely reflected and no light is transmitted through the prism.

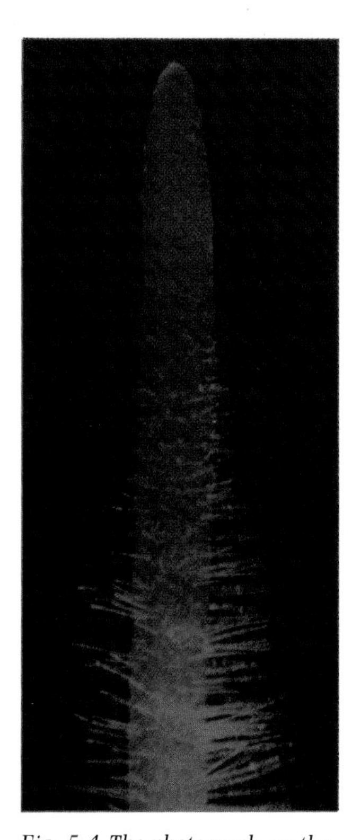

Fig. 5.4 The photograph on the right shows an optical fibre wound round a drum carrying light from a helium–neon laser. The fibre is about 100 m long and has been deliberately badly made so that some light leaks out of the sides so that we see the fibre as red. In a high quality fibre nearly all the light would emerge from the end. In this case, the light coming from the end of the fibre is directed onto a screen. The photograph above shows red laser light guided through the tissue of a corn root seedling. Tiny root 'hairs' pipe the light out to the tips of the hairs.

without violating the conservation of energy, in quantum mechanics, if the time uncertainty is Δt, we cannot know the energy to better than an uncertainty $\Delta E = h/\Delta t$. Roughly speaking then, we can 'borrow' an energy ΔE to get over the barrier so long as we repay it within a time $\Delta t = h/\Delta E$. If the barrier is too high or too wide, therefore, tunnelling becomes extremely unlikely and all the electrons will be reflected, just like the roller coaster car. Needless to say, this sort of 'hand-waving' argument must be backed up by detailed calculation with the Schrödinger equation, but such arguments do give us some sort of insight into quantum tunnelling. It is perhaps more illuminating, however, if we look at the behaviour of more familiar waves. This phenomenon of tunnelling is then seen to be a general property of wave motion – it only becomes surprising when taken in conjunction with de Broglie's hypothesis that all quantum 'particles' have wavelike properties.

Wave tunnelling

Although both waves on a string and water waves can be made to exhibit 'wave tunnelling', probably the most familiar example involves light in its wavelike guise. Consider what happens when light travels from air into a block of glass. As shown in fig. 5.3, because light travels slower in glass than in air, the wave slews round and the light changes direction. This is called 'refraction'. Now consider light travelling from glass to air. Instead of being bent towards the vertical, the light is bent away from it. If we increase the angle at which we shine the light on the glass–air surface, there will be an angle – the 'critical' angle – at which the light emerges in the air just grazing

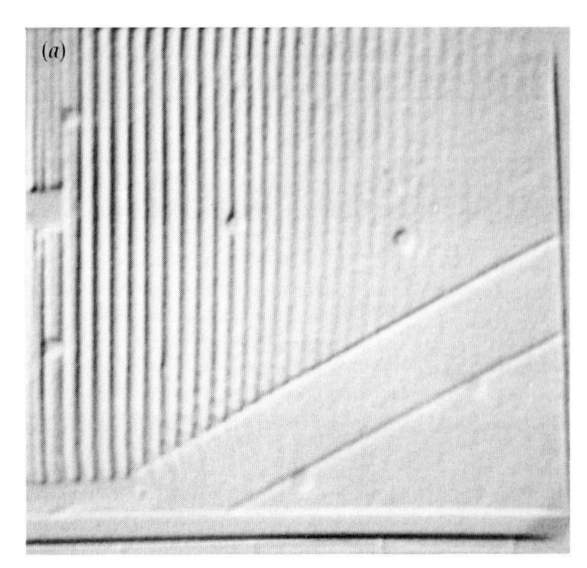

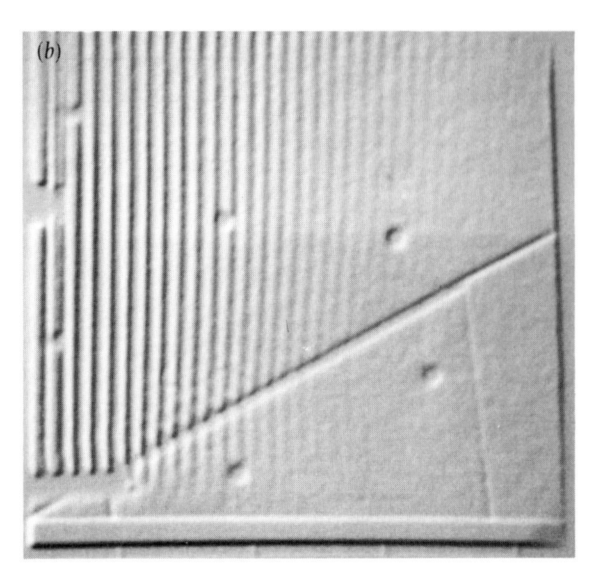

Fig. 5.5 Tunnelling with water waves. (a) *The speed of these water waves depends on the depth of water. This photograph shows the water waves undergoing 'total internal reflection' at a barrier consisting of a change in water depth. Notice that there is some kind of disturbance in the forbidden region beyond the barrier but that it does not correspond to an ordinary wave.* (b) *This photograph shows that as the width of the forbidden region is decreased, the water wave can 'jump the gap' and appear on the other side. This standard wave phenomenon is the basis for tunnelling in quantum mechanics.*

the surface. What happens if we increase the angle still further? What must happen is that all the light is reflected from the glass–air surface and no light escapes into the air. This is called 'total internal reflection' and is the basis for modern fibre optics. How is all this connected with quantum tunnelling? Well, although no light rays penetrate the air beyond the glass when the light arrives at an angle larger than the critical angle, there is nonetheless some sort of wave disturbance in the air. This is not a wave that carries energy, like ordinary 'travelling' waves, but a sort of 'standing' wave pattern that does not transmit any light energy. The wave patterns on a string fixed at both ends are examples of standing waves. However, the type of standing wave involved here – a so-called 'evanescent' wave – is special in that the disturbance dies away very rapidly the further away we go from the surface. The connection with tunnelling comes about if we bring up another block of glass parallel to the first one. As we bring the two blocks towards each other,

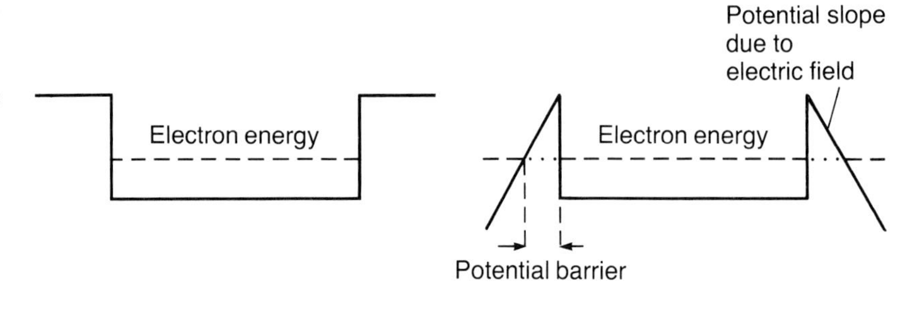

Fig. 5.6 The diagram on the left shows a simplified picture of the potential well for electrons in a metal. The broken line represents the electron energy which is insufficient to escape from the well. On the right, however, we show how the potential is modified in the presence of a large electric field. Electrons can now escape from the metal by tunnelling through the barrier.

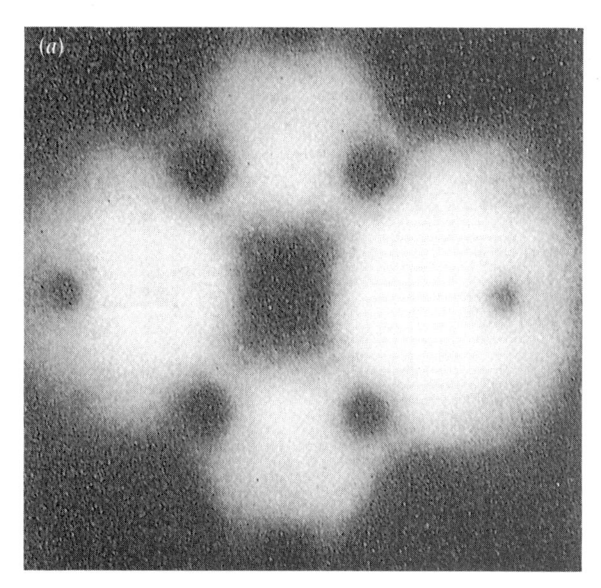

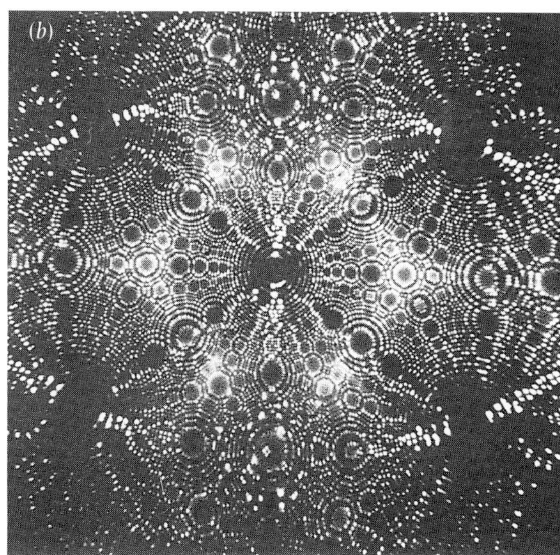

Fig. 5.7 (a) *The left hand picture is the tip of a tungsten needle photographed with an electron field-emission microscope. The needle is negatively charged so that there is a very large electric field at the tip. Electrons tunnel out and are accelerated away from the needle by the high field. This produces a very high magnification of the needle showing where the electrons are most densely clustered. The resolution is not quite high enough to see individual atoms: the bright areas in the photograph correspond to electron emission from the atoms at the corners and edges of the layers making up the tip of the needle.* (b) *The right hand picture is a helium ion emission microscope photograph of the tip of a tungsten needle. In this case, the needle is positively charged and surrounded by helium gas. Electrons are captured by the high fields at the tip of the needle and the remaining positively charged helium ions accelerated away. The helium ions are heavier than the electrons used above and suffer less from random sideways motion as well as having a smaller quantum wavelength. These two improvements combine to improve the resolution sufficiently so that individual atoms may be seen. Each bright spot on the picture corresponds to a tungsten atom. The pattern of rings in the image can be understood by imagining the atoms in a metal to be packed in regular layers, like apples in a box. If we now imagine cutting a cone shape from the metal to form the tip of a needle, there will be circular rings as each layer is exposed.*

the evanascent wave begins to penetrate the second block and a transmitted ray of light appears! The closer the two blocks are brought together, the more light energy that reappears as a transmitted ray. This is because the amplitude of the standing wave in the 'forbidden' air gap has not had time to decay away so much. Physicists call this phenomenon 'frustrated total internal reflection' and it is an exact analogue of quantum barrier penetration or tunnelling for de Broglie waves. Fig. 5.5 shows a ripple tank photograph of barrier penetration with water waves. This phenomenon is also useful in modern optics as a 'beam splitter': the amount of light transmitted can be controlled by adjusting the width of the forbidden gap.

Applications of quantum tunnelling

There are many devices now in common use that rely on the ability of quantum particles to tunnel through barriers. The two examples we shall describe here both involve electrons, but there are other examples that we shall come across later in which alpha particles and pairs of electrons are doing the tunnelling. Our first example concerns the so-called 'field-emission' microscope. In a metal, the electrons that carry the electric currents are able to move about relatively freely. As a very simplified model of a metal, the electrons can be imagined as moving about in an attractive potential well, rather like the quantum 'box potential' we talked about in the previous chapter. Since it takes energy to knock electrons out of the metal, there must be an electrical 'hill' or barrier at the sides that keeps them in. If we now turn on a strong electric field, there is a large attractive force that wants to drag electrons out of the metal. The resulting potential looks like fig. 5.6*b*. As can be seen, there is now a barrier that the electrons inside the metal can tunnel through and escape. Using the Schrödinger equation we can calculate the probability for this to happen. One finds that a very large electric field is needed for there to be appreciable tunnelling. Such large fields can be produced at the tip of a sharp needle of metal, and experiments in 1928 confirmed the predictions of quantum mechanics for this 'field-emission' process. Several years later, this effect was used as the basis for a new form of microscope. After tunnelling through the barrier, the electrons will move away from the needle in a straight line. Thus, if we surround the needle by a phosphorescent screen we will see a greatly enlarged image of the tip of the needle caused by the electrons arriving at the screen. Electrons are most easily emitted from the corners and edges of layers of atoms in the tip, and these positions show up brightest in the image. Magnifications of up to one million can be reached in this way.

An even more sensitive device was later developed from this field-emission microscope. Instead of electrons arriving at the screen, the 'field-ion' microscope uses helium ions. This comes about as follows. Instead of a vacuum surrounding the needle, a small amount of helium gas is allowed in and, in addition, the field on the needle is reversed. When a helium atom collides with the tip, the high electric field not only ionizes the helium but accelerates the positively charged ion away from the needle. Because the helium ions are much more massive than electrons (about 8000 times heavier) they have much larger momentum and a correspondingly shorter de Broglie wavelength. Thus, this 'field-ion' microscope is able to pick out much finer details of the structure of the tip of the needle. The bright spots on the image reveal the positions of individual metal atoms on the surface.

A new type of microscope that relies on quantum tunnelling has recently been developed. This is the so-called 'scanning tunnelling microscope', which enables one to achieve magnifications of up to 100 million and to reconstruct the surfaces of solids atom by atom. The basic idea is very simple. According to quantum mechanics, electrons in a solid have a small but non-vanishing chance of being found just outside the metal surface. The probability for this to happen falls off very rapidly with distance away from the surface. Now, if a needle-like probe is brought up very close to the surface, and an electrical voltage applied between it and the metal, a tunnelling current will flow across the gap between them. Since the magnitude of this current is extremely sensitive to the distance of the probe from the metal, it is possible to scan across the metal systematically and to reconstruct a very accurate contour map of the surface. This type of microscope is much more versatile than the field-ion microscope and promises to have many new applications in physics, chemistry and biology.

Our second example of barrier penetration by electrons is a device used in modern electronics called the 'tunnel diode'. This consists of a junction between two types of semiconductor to which a voltage can be applied. We shall explain how semiconductors work in the next chapter – here all we need to know is that the electrons carrying the current 'see' a potential barrier like a wall. Electrons can tunnel through the barrier from one side of the junction to the other, but the amount of this 'tunnelling' current depends very sensitively on the height of the barrier. One can therefore make a very rapid 'switching' device in which the amount of current allowed to flow through the device is varied by changing the applied voltage and thus altering the height of the barrier.

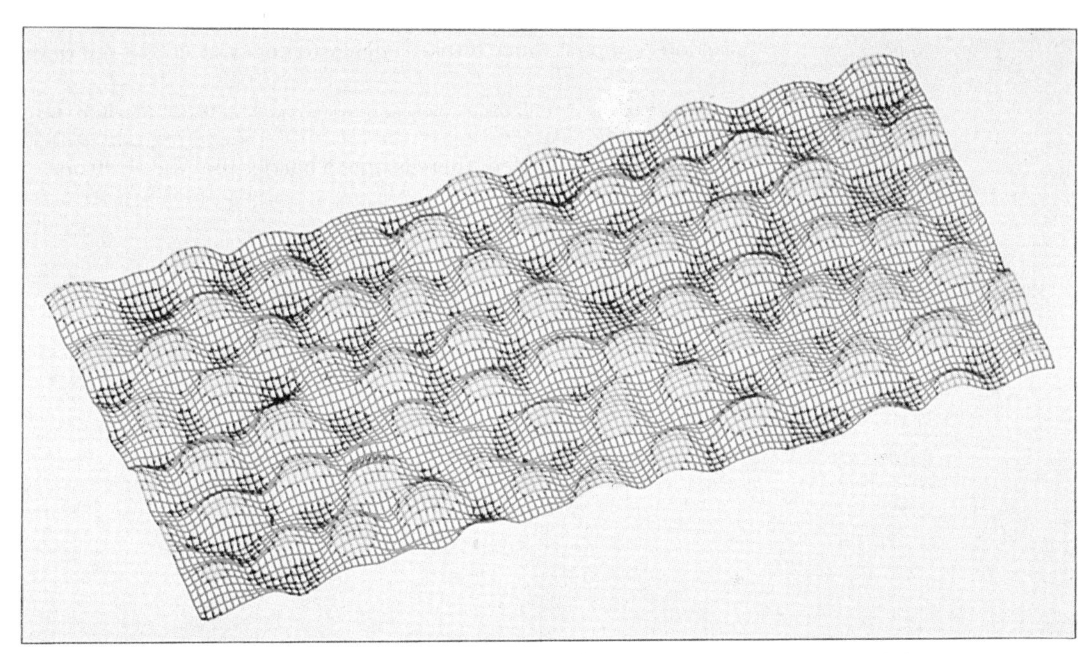

Fig. 5.8 A scanning tunnelling microscope consists of a very sharp needle point that surveys the surface of a specimen very accurately. When a high voltage is applied between the surface and the needle, electrons can tunnel from the tip of the probe to the specimen under investigation. This tunnelling current is very sensitive to the height of the probe above the surface. In the microscope, the height of the needle can be adjusted as the needle moves over the surface so that the current remains constant. In this way, the up and down movements of the needle map out the detailed contours of the surface. The colour coded picture shows the surface of silicon with each of the yellow bumps representing a silicon atom. A regular diamond structure consisting of 12 atoms of silicon may be picked out.

Nuclear physics and alpha decay

One of the great puzzles in the early days of nuclear physics concerned alpha decay. The puzzle was this. In the radioactive decay of uranium, physicists had measured the energy of the alpha particle that was thrown out of the nucleus and found it to be about 4 MeV. A quick word about units of energy is needed here. An electron-volt, or 'eV', is the amount of energy an electron gains in 'sliding down' a potential hill one volt high. This amount of energy is typical of the energies of electronic energy levels in atoms. For processes involving the nucleus, on the other hand, the energies are much larger, and a convenient unit of energy is a million electron-volts, or 'MeV' for short. Now back to the story. Rutherford had done experiments shooting alpha particles at atoms and had found that alpha particles with about 9 MeV of energy were repelled by the positive charge of the nucleus. In other words, to get inside the nucleus requires much more energy than the 4 MeV observed for alpha particles emitted in radioactive decay. To make this problem more

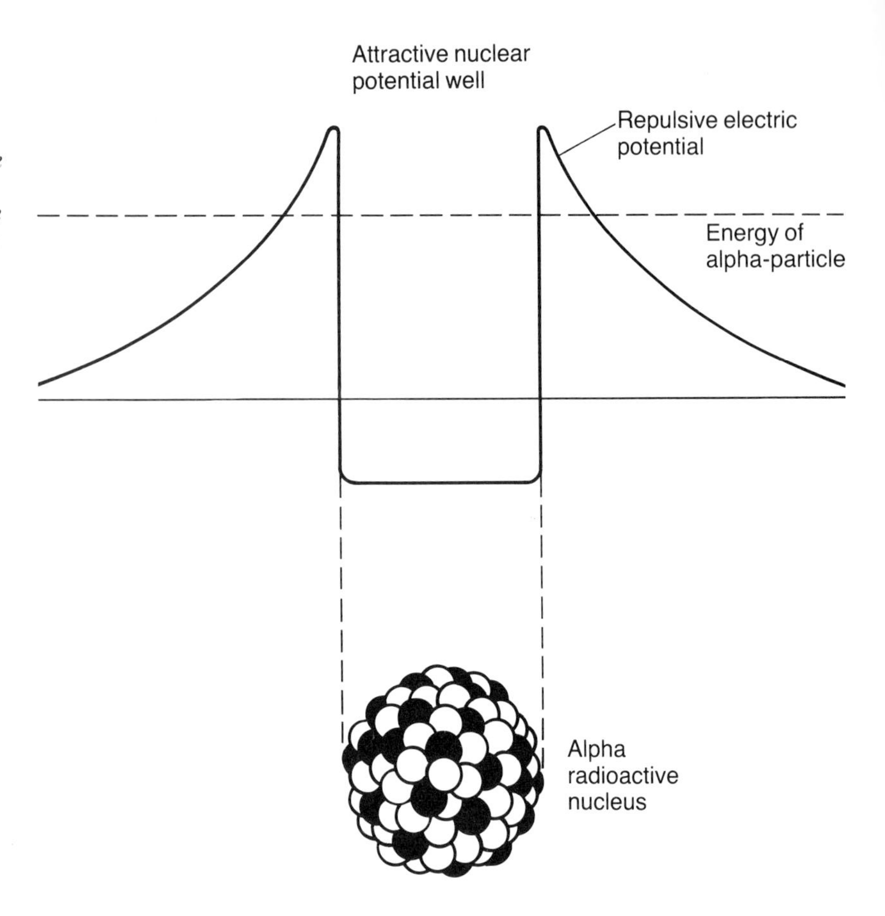

Fig. 5.9 Gamow's original application of quantum tunnelling was to the process of alpha decay. Because alpha particles are unusually stable we can think of them as existing in a nuclear potential due to all the other particles in the nucleus. It is clear that it is possible for an alpha particle to tunnel out of the nucleus causing it to decay.

George Gamow (1904–68) was the creator of the intrepid explorer Mr Tompkins. Gamow was the grandson of a Tsarist general and obtained his PhD from Leningrad University. After working at most of the important scientific centres in Europe, he eventually settled in the USA. Besides being one of the best-known popularizers of science, Gamow made important contributions to nuclear physics.

graphic, let us look at the analogous situation for our roller coaster. It is as if you were sitting on the track, half-way down the hill, and were suddenly bumped into by a carriage. The only place from which the carriage could have come was from over the top of the hill. But if the carriage had indeed come rocketing down from the top of the hill it would have caused you a nasty injury. Instead, it hit you with just a light bump!

In view of what we have been saying about tunnelling, the answer to the alpha particle paradox is now fairly obvious. But in 1928, when an explanation of alpha decay in terms of quantum tunnelling was suggested by the Russian physicist George Gamow, and by two US physicists, Edward Condon and Ronald Gurney, it was a startlingly new idea and one of the first applications of quantum mechanics to the nucleus. In the nucleus of the common isotope of uranium, ^{238}U, there are 92 protons and 146 neutrons and these are all jostling about in a very small volume. The strong nuclear forces between the 'nucleons' – protons and neutrons – can be thought of as providing an attractive 'potential well' that keeps them inside the nucleus, rather like our simple model for conduction electrons in a metal. However, inside the nucleus two protons and two neutrons can sometimes get together to form an alpha particle. The resulting potential 'seen' by the alpha particle is shown in fig. 5.9. Although the height of the barrier is around 30 MeV, the alpha particle can tunnel its way out of the nucleus and appear as a free particle with only 4 MeV of energy. We now know much more about nuclear forces and can perform calculations using more realistic nuclear potentials. Nonetheless this simple picture of alpha decay still works remarkably well.

A photograph of Cockcroft (right) and Walton (left), on either side of Rutherford.

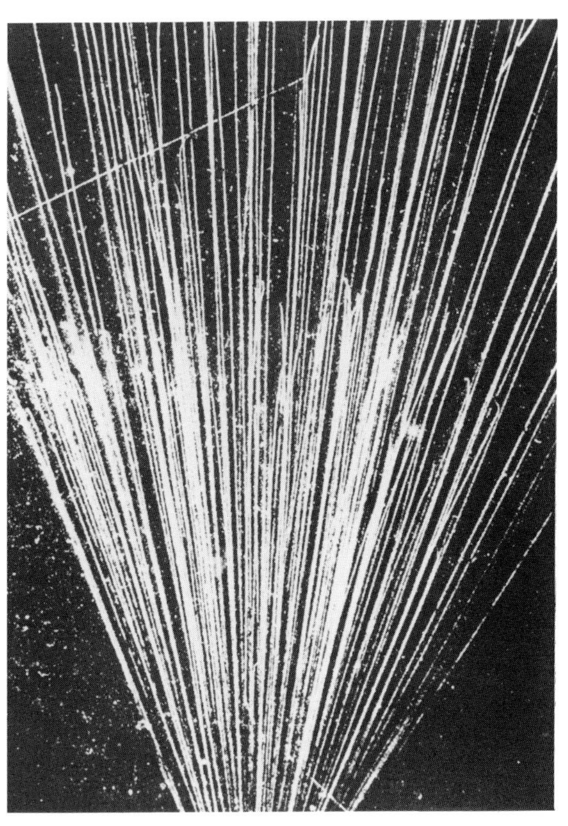

Fig. 5.10 Tracks left by alpha particles in a cloud chamber. The tracks consist of tiny droplets of water which have condensed in the wake of an alpha particle. Since the chance of an alpha particle encountering the nucleus of one of the gas atoms is very small, the tracks are almost straight. At the top of the picture, however, there is a backward going track and a small kink in the track of the alpha particle. This is, in fact, a nuclear reaction: the backward going track is a proton produced in the collision of the helium nucleus with the nitrogen nucleus. The short forward going track is left by an oxygen nucleus created in the nuclear reaction.

There is an interesting 'converse' to this problem that resulted in Cockcroft and Walton winning the Nobel Prize. In 1919, Rutherford, again in his experiments with alpha particles, had observed the first artificially produced nuclear reaction. On firing alpha particles at nitrogen he had observed that protons were occasionally produced. Rutherford deduced that he had seen a 'nuclear disintegration' and we would now write this reaction as

$^{4}_{2}He$	+	$^{14}_{7}N$	$\rightarrow$	$^{17}_{8}O$	+	$^{1}_{1}H$
helium		nitrogen		oxygen		hydrogen
2 protons		7 protons		8 protons		1 proton
2 neutrons		7 neutrons		9 neutrons		0 neutrons

Fig. 5.10 shows an early photograph of this reaction. Rutherford concluded his scientific paper on this observation with a remark that 'If alpha particles – or similar projectiles – of still greater energy were available for experiment, we might expect to break down the nucleus structure of many of the lighter atoms'. This possibility was realized with the coming of new types of particle accelerators. In 1932, the same year that the neutron was discovered by Chadwick, a US physicist called Ernest Lawrence had built a machine called a 'cyclotron', which could accelerate particles to energies well over 1 MeV. At the time, however, it was widely believed that in order for charged particles to penetrate to the core of the nucleus, the projectiles would have to have energies of many MeV. In fact, Cockcroft and Walton, working in Cambridge, UK, with a much more primitive accelerator, were the first to 'split the atom' using artificially accelerated protons with an energy of less than 1 MeV! It is said that Cockcroft, in a rare and atypically flamboyant gesture, was seen wandering through the streets of Cambridge announcing 'We've split the atom!' to all and sundry. In actual fact, the name 'atom

Fig. 5.11 The electrostatic generator of Cockcroft and Walton. Cockcroft is sitting in the cage under the apparatus.

Livingstone and Lawrence next to their cyclotron. This was used in 1937 to produce the first artificial element, technetium. Technetium has 43 protons but does not occur naturally because all its isotopes are radioactive with short lifetimes.

Fig. 5.12 High voltage generator used as a pre-injector for particles in the NIMROD accelerator at the Rutherford Laboratory, near Harwell, UK.

smashers' belongs to the popular press. What Cockcroft and Walton had observed was the first artificially induced nuclear reaction – a modern version of the old dreams of alchemists about transmutation of the elements. They had converted lithium into helium via the reaction

$$^1_1\text{H} + ^7_3\text{Li} \rightarrow ^4_2\text{He} + ^4_2\text{He}$$

Their experiment could have been performed by Lawrence almost a year before but had not been thought worthwhile to attempt, because it was thought necessary to have enough energy to surmount the electric repulsive barrier surrounding the nucleus. Lawrence was on honeymoon on a boat in

Connecticut at the time that the Cockcroft and Walton result hit the headlines. At once he sent a telegram to his colleague James Brady back in Berkeley, California: 'Cockcroft and Walton have disintegrated the lithium atom. Get lithium from chemistry department and start preparations to repeat with cyclotron. Will be back shortly'. Brady showed the telegram to his fiancée with the comment 'That's what physicists on their honeymoon think about'.

There is an important postscript to the Cockcroft and Walton experiment. They were able to predict the energies of the two alpha particles they observed in the reaction using energy conservation only if they assumed Einstein's famous relation between mass and energy

$$E = mc^2$$

$$\text{energy} = \text{mass times (velocity of light)}^2$$

This relation tells us that mass may be regarded as just another form of energy, and that the amount of energy, E, that is equivalent to any given mass, m, may be calculated from the formula above. Thus, for the reaction they had observed they added up the masses of the particles on both sides of the reaction equation, together with all the kinetic energies, and obtained a 'mass–energy' equation of the form

proton mass + lithium mass = 2 × (helium mass + kinetic energy of helium)
+ kinetic energy of proton

The kinetic energy predicted for the two helium nuclei was about 8.5 MeV, much larger than the initial proton kinetic energy, and in good agreement with the experimental measurements. This principle of mass–energy balance is basic to the application of energy conservation in the whole of nuclear physics.

Nuclear fusion and nuclear fission

The mass–energy relation used by Cockcroft and Walton to predict the energies of the two alpha particles in the first artificial nuclear reaction leads directly to an understanding of the 'binding energies' of nuclei. The simplest nucleus after ordinary hydrogen is that of deuterium, which contains one proton and one neutron. Why do these particles stay together? The reason is that there is an attractive 'strong' nuclear force that binds them together so that they have a lower energy, when combined as a deuterium nucleus (the deuteron), than when they exist as separate particles. We can calculate this binding energy, B, using the same sort of mass–energy equation as before:

$$B = m_p c^2 + m_n c^2 - m_D c^2$$

binding energy = mass energy of proton + mass energy of neutron − mass energy of deuteron

Using the experimentally measured mass values we find that the binding energy is about 2 MeV. This is the energy that would be liberated if we could take a proton and a neutron and put them together to make a deuteron. By measuring the masses of all the different nuclei we can calculate the binding energy of each nucleus. If we label the number of protons in a nucleus by Z and the number of neutrons by N we say that the total number of 'nucleons', A, is just the sum of these, i.e.

$$A = Z + N$$

In fig. 5.13 we show a plot of the 'average binding energy per nucleon' for all the various elements. We see that the binding energy rises from about 2 MeV, the value we have just calculated for the deuteron, up to a maximum of

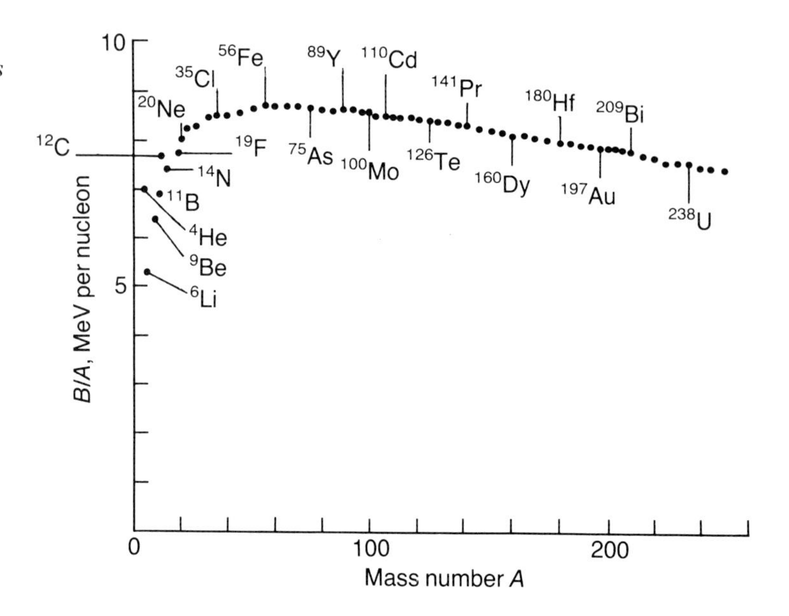

Fig. 5.13 A plot of the 'binding energy per nucleon' versus 'mass number A', the total number of nucleons in the nucleus. The word 'nucleon' means both protons and neutrons. Roughly speaking, the binding energy is the energy required to remove a nucleon from the nucleus. This figure shows that iron is the most stable nucleus and that helium is much more strongly bound than other elements nearby.

around 8.8 MeV per nucleon for iron (Fe) and then falls gradually to about 7.5 MeV for heavy nuclei out to uranium. Notice that alpha particles (helium nuclei) are especially stable compared with the elements nearby. This is why they are sometimes formed inside heavy nuclei and can tunnel out, causing radioactive decay of the nucleus. The position of iron as the most strongly bound nucleus shows that there are two ways of releasing energy from the nucleus. One is the process of 'fusion', in which two nuclei lighter than iron combine to form a heavier nucleus, and the other is 'fission', in which a very heavy nucleus splits into two lighter ones. The binding energy released in these two processes appears as kinetic energy of the final particles.

Let us look at an explicit example of a fusion reaction. By looking at all the different binding energies one can find many possible reactions. Perhaps the best candidate for a fusion nuclear power reactor is the so-called 'D–T' or deuterium–tritium reaction. This uses these rare isotopes of hydrogen in the reaction

$$^{2}\mathrm{H} + {}^{3}\mathrm{H} \rightarrow {}^{4}\mathrm{He} + \mathrm{n}$$

$$\text{deuterium} + \text{tritium} \rightarrow \text{helium} + \text{neutron}$$

to obtain an energy release of 17.6 MeV. The problem with generating power in this way is that it is difficult to produce an environment in which a sustained reaction is possible. Why is this? The answer is, because of the electrical repulsive barrier experienced by the particles as they approach each other. While it is easy to produce this reaction using beams of deuterons accelerated to energies greater than this 'Coulomb' barrier, this is not a viable way of producing large amounts of power commercially. Instead, the approach being followed in the search for cheap fusion power is that of heating up the initial constituents to a very high temperature so that there is enough kinetic energy in ordinary collisions of the hot gas or 'plasma' to enable this reaction to take place. Generation of such high temperatures and 'containment' of the hot plasma present formidable technical obstacles which will not be overcome for many years. It may therefore be somewhat surprising that such fusion reactions are the basic process by which the stars generate their energy, despite the fact that the temperatures in stars lead to kinetic energies much lower than the Coulomb barrier! In fact, fusion is able to take place at these low temperatures only

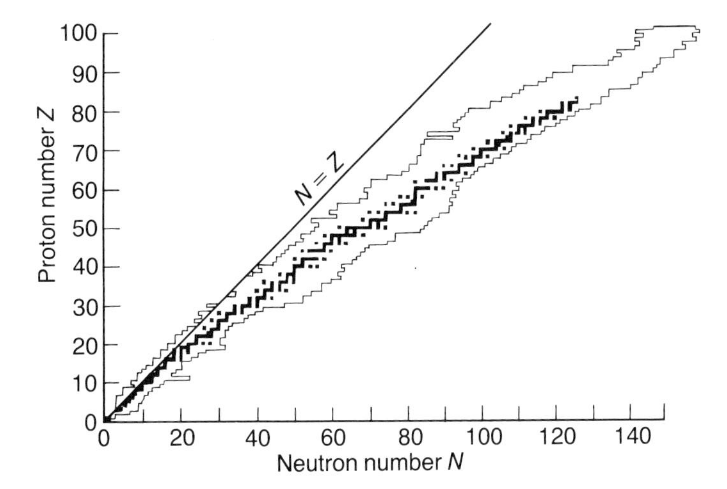

Fig. 5.14 A compilation of all observed nuclei plotted in terms of the number of protons in the nucleus versus the number of neutrons. Stable nuclei are shown in black and unstable nuclei in light shading. More massive nuclei contain more neutrons than protons. The protons are positively charged and produce an electric repulsive energy which tends to make the nucleus unstable.

because of quantum tunnelling through the potential barrier. Thus, it is not an exaggeration to say that we owe our very existence to the ability of quantum particles to penetrate classically forbidden regions!

Radioactivity is a process by which a nucleus with Z protons and N neutrons can transform itself to one with a different Z and N. Many nuclei are stable and do not decay at all. A plot of all known stable nuclei is shown in fig. 5.14. together with radioactive nuclei shown in lighter shading. The three types of 'radiation' emitted by radioactive nuclei were historically called alpha, beta and gamma rays. Alpha particles are helium nuclei and so alpha decay decreases both Z and N by two. In contrast, the basic beta decay process does not change the value of A but involves a neutron changing into a proton with the emission of an electron. It is also possible for a proton to change into a neutron and for an 'antielectron' or 'positron' to be ejected. We shall talk about antimatter and beta decay processes in more detail in a later chapter. Finally, the mysterious gamma rays so beloved of comic strips are nothing more than very energetic photons. They arise because some other radioactive decay process – alpha or beta decay – has left the new nucleus in an 'excited' state. These excited states of nuclei are similar to those of the hydrogen atom apart from the energy differences between the energy levels. For atoms, the energy level differences and photon energies are typically measured in units of electron-volts (eV): for nuclei, the corresponding energy differences and photon energies involve millions of electron-volts (MeV).

Why are some nuclei stable and others radioactive? To understand this in detail clearly requires a detailed picture of nuclear forces. However, some general features can be understood rather easily, from the fact that the very strong attractive nuclear forces are only effective over very short distances, typically smaller than the size of a heavy nucleus. This is why we see no effect of this enormously strong force on the scale of everyday objects. The electric force, on the other hand, is much weaker, but is effective over much larger distances, much larger than the size of the nucleus. Amongst the neutrons and protons in a nucleus there is therefore competition between the short range nuclear force that wants to hold the nucleons together and the electrical repulsion of the protons which is trying to break the nucleus apart. For most nuclei, the nuclear force wins out, but for heavy nuclei there is a delicate balance between these two opposing forces. If we keep adding protons and neutrons to make heavier and heavier nuclei, the range of the electric force is large enough for all the protons to repel each other. On the other hand, the nucleus is now so big, and the nuclear force so short ranged, that any given nucleon only feels the strong attractive nuclear forces from

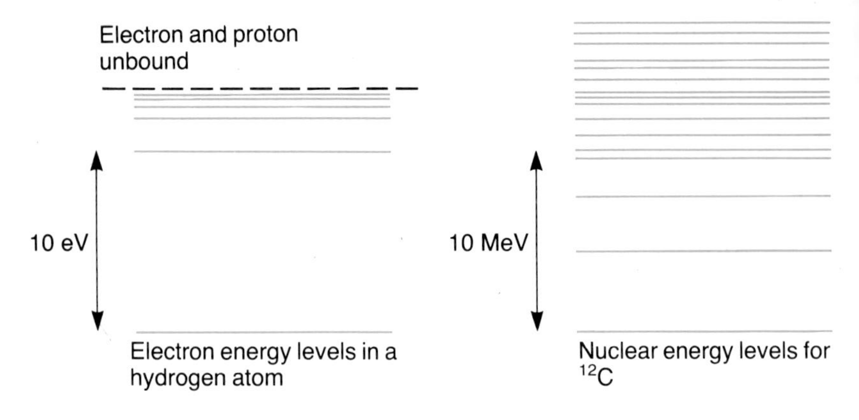

Fig. 5.15 A comparison of atomic and nuclear energy levels for hydrogen and ^{12}C. The nuclear binding energy is about a million times larger than that binding an electron to an atom. When atomic electrons make a transition from one level to another the energy of the photons typically corresponds to that of visible light. In a nuclear transition, the very high energy photons emitted are usually called gamma rays.

nearby nucleons. This is the reason why stable heavy nuclei with very large *A* values have more neutrons than protons: the excess neutrons give more attractive binding energy without any Coulomb repulsion. However, if we take an unstable large *A* nucleus and change one of the protons into a neutron by beta decay, or eject an alpha particle, we end up with a new nucleus which has fewer protons and less Coulomb repulsion, and is therefore more strongly bound. This is the essence of nuclear stability.

This picture of the nucleus also shows that there is the possibility of a heavy nucleus lowering its energy in a more spectacular way. Bohr had suggested that a heavy nucleus should be imagined as a sort of 'liquid drop' rather than a brittle solid. Furthermore, in a heavy nucleus the balance between Coulomb repulsion and attractive nuclear forces is very delicate. Perhaps if the nucleus is disturbed by firing another neutron into it, the 'drop' may be able to break up into two smaller drops. Looking at the binding

Fig. 5.16 Schematic representation of nuclear fission and its relation to quantum tunnelling.

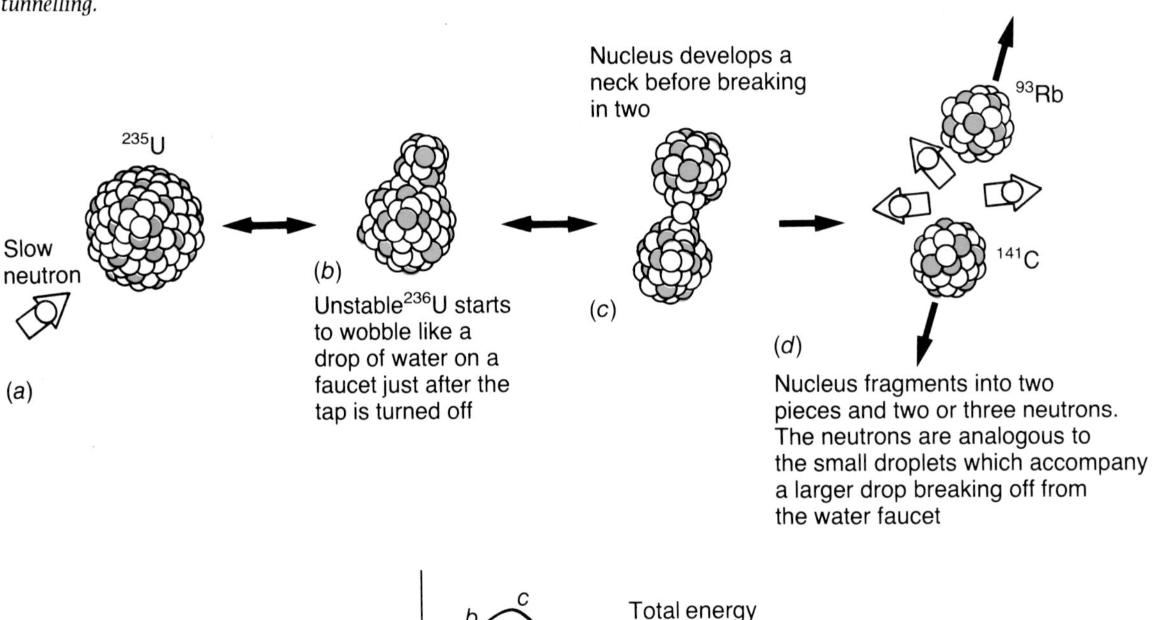

Lise Meitner and Otto Hahn in their laboratory in Berlin in 1920. After working with Hahn for over 20 years, Meitner left Germany in 1938 when Austria was occupied by Hitler. Her faith in the correctness of Hahn's surprising results led to the discovery of nuclear fission.

energy curve (fig. 5.13) you can see that we may be able to gain energy in this way if the neutrons and protons in the two lighter nuclei have less mass-energy than when they are all together in one very heavy nucleus. Such a possibility was first suggested by a woman chemist called Ida Noddack who was criticizing some experiments done by Fermi and his group in Rome. This group had bombarded uranium with neutrons and thought they had produced new 'transuranic' elements with Z greater than the 92 of uranium. Noddack had objected that they had not proved that uranium did not split instead into two large fragments. The objection was not followed up, however, and the famous German chemist Otto Hahn was delighted to have lots of new transuranic elements to study! To everyone's amazement, however, Hahn and his pupil Fritz Strassman were reluctantly forced to conclude that instead of a new element they had only found some isotopes of barium with a Z of 56. This discovery took place at the end of 1938, on the eve of the Second World War. Because of the prevailing hostility towards Jews, Hahn's collaborator of 30 years, Lise Meitner, whom Einstein once called the German Madame Curie, had been forced to flee to Sweden. It was over Christmas of that year, with her nephew Otto Frisch, that she learnt from Hahn that he had found barium in the reaction products. Lise Meitner was convinced that Hahn was too good a chemist to have made a mistake, and it was on a walk in the woods in the snow that she and Frisch arrived at the answer. Their paper was written by long-distance telephone calls between Stockholm and Copenhagen some days later, and crucial experiments to confirm their idea were performed by Frisch in just two days! The name 'fission' was coined by Frisch from the name biologists use to describe the process by which single cells divide into two.

This process of nuclear fission can also be thought of as a tunnelling process, similar to the others we have been talking about in this chapter. The energy of the fissioning nucleus can be pictured as a roller coaster potential like that of fig. 5.1. There are two valleys or minima of energy but one is lower than the other. Classically, a particle in the upper valley will stay there for ever. Quantum mechanically such a state is not completely stable – the system has the possibility of tunnelling through to the true lowest energy

Fig. 5.17 A painting of the opening ceremony of the world's first nuclear reactor. Fermi's reactor used a controlled and sustained nuclear reaction involving uranium.

state. The 'false' minimum is known as a 'metastable' state and fission may be imagined as such a tunnelling process.

There are two postscripts to this discovery of nuclear fission. One concerns the discovery of true transuranic elements. These were discovered in early 1940 using Lawrence's cyclotron by Ed McMillan. Since the planets Neptune and Pluto lie beyond the planet Uranus, these first transuranic elements, with Z of 93 and 94, respectively, were called neptunium and plutonium. Both were unstable, but one of them, plutonium, was later to have sinister implications in the production of nuclear weapons.

How nuclear weapons came about brings us to our second postscript. Nuclear reactions, in which the energy liberated by mass conversion is about 100 million times larger than that available in chemical reactions, clearly have enormous potential both as an energy source or as a weapon. Moreover, one vital point about fission that was missed by Meitner and Frisch was the possibility of a 'chain reaction'. A typical fission reaction liberates a couple of free neutrons in addition to the two large fission fragments:

$$^{235}_{92}U + n \rightarrow ^{93}_{37}Rb + ^{141}_{55}Cs + 2n$$

The rare isotope of uranium, ^{235}U, fissions into rubidium and caesium plus two extra neutrons. The barium observed by Hahn and Strassmann comes from radioactive decay of the unstable caesium isotope. Now, each of the neutrons produced can initiate another fission reaction, and the resulting neutrons produced by these reactions generate more fissions until an avalanche of fissions or a 'chain reaction' is set up. This chain reaction can either release energy in a controlled manner, as in a nuclear reactor where the number of neutrons causing fissions can be regulated, or in a catastrophic explosion, as in a fission nuclear bomb. Fortunately, to make a chain reaction go you need to have a lot of fissionable material, and only the rare isotope ^{235}U is naturally occurring and suitable for either a reactor or a bomb. However, plutonium, which could be produced by a nuclear reaction involving the common isotope of uranium, ^{238}U, is also suitable. The first 'atom bomb' ever tested was a plutonium bomb, and 'Fat Man' dropped on Nagasaki on August 9th, 1945, was also made from plutonium. 'Little Boy' dropped on Hiroshima on August 6th, 1945, was a uranium bomb.

The development of a reactor with a controlled fission chain reaction was vital for the production of plutonium during the Second World War. The first man-made nuclear reactor was built in Chicago by Fermi in 1942, but Nature had done it first. It is believed that a natural fission reactor operated two thousand million years ago in Africa using natural uranium deposits!

LAST EDITION

HP sauce

Daily Mail

TUESday FIELD-DAY

NO. 15,367 ONE PENNY FOR KING AND EMPIRE TUESDAY, AUGUST 7, 1945

Most terrifying weapon in history: Churchill's warning

'THIS revelation of the secrets of nature, long mercifully withheld from man, should arouse the most solemn reflections in the mind and conscience of every human being capable of comprehension. We must indeed pray that these awful agencies will be made to conduce to peace among the nations, and that instead of wreaking measureless havoc upon the entire globe they may become a perennial fountain of world prosperity.' SEE BELOW.

ATOMIC BOMB: JAPS GIVEN 48 HOURS TO SURRENDER

Radios threaten Tokio: 'You can expect annihilation'

ZERO HOUR CAME

CHURCHILL TELLS BRITAIN'S PART

Spies, RAF, commandos

Fig. 5.18 Newspaper headline announcing the atomic bomb ultimatum given to the Japanese.

Fig. 5.19 A soldier on picket duty at Nagasaki was vaporized by the explosion even though he was 3.5 km from the centre of the blast.

Fig. 5.20 A photograph of an atomic artillery test performed in the Nevada Desert in 1953. The weapon was fired from a 280 mm gun, and was slightly more powerful than the bomb that destroyed Hiroshima.

Fig. 5.21 Enola Gay, *the B-29 bomber that dropped the first atomic bomb on Hiroshima, is now immortalized in a recent hit record.*

Fig. 5.22 An awe-inspiring photograph of the mushroom cloud from a nuclear test explosion.

Fig. 5.23 Possible effect of a one-megaton nuclear explosion on Manhattan. The explosion could result in a massive firestorm engulfing much of the city. Furthermore, the smoke and dust thrown up into the atmosphere in a full-scale nuclear war could result in a very severe 'nuclear winter' with drastically reduced surface temperatures.

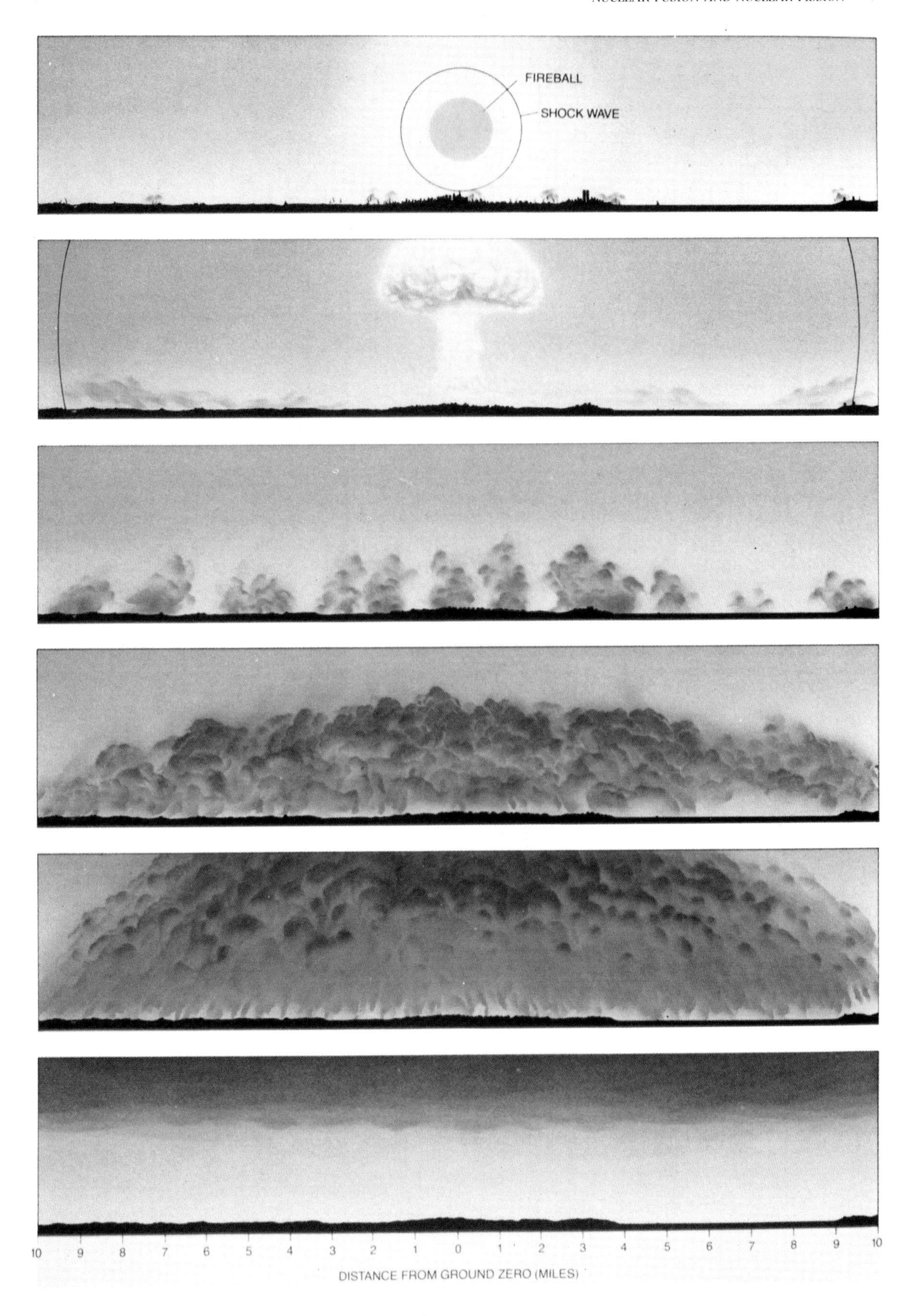
FIREBALL
SHOCK WAVE
10 9 8 7 6 5 4 3 2 1 0 1 2 3 4 5 6 7 8 9 10
DISTANCE FROM GROUND ZERO (MILES)

Dating using radioactivity

Almost everything around us is slightly radioactive. The air we breathe, the soil in our gardens, most building materials and even our own bodies, all contain radioactive elements. Much of this radioactivity originates from naturally occurring uranium and thorium. It may come as a surprise to you to know that*, 'on the average over the Earth's crust, each square mile has, within one foot of the surface, 8 tons of uranium and 12 tons of thorium'. Both uranium and thorium form one end of complicated decay chains of radioactive elements which eventually end up with stable isotopes of lead. On the way to lead, however, these chains produce radioactive gases which are isotopes of radon. In the lungs, these gases are very dangerous and can cause cancer. It is said* 'that every miner who has worked in the Joachimstal uranium mines in Czechoslovakia for more than ten years has died of lung cancer'. Modern mines now have powerful ventilation systems to flush away these gases. This is one reason for us not to make our houses too draught-proof!

Given a sample containing a large number of identical radioactive nuclei, we cannot predict when any particular nucleus will decay. However, by measuring the number of decays that occur in the sample in a given time, we can calculate the probability that a nucleus will decay in the next second. We often quantify this probability in terms of a 'half-life' – this is the time taken for half the sample to decay into other elements. Values for half-lives vary over an enormous range: from 4500 million years for the ^{238}U, through 1600 years for radium, 3.8 days for radon, to much less than a second for polonium. What this means is that if some material was formed 4.5 thousand million years ago from pure uranium, there would now only be half as much uranium and the rest would have transformed into lead. By measuring the relative amounts of different isotopes in small samples of rock, it is possible to estimate the age of the rock. This technique has been used to date Moon rocks and meteorites as well as terrestrial rocks. Using samples of rock from many different parts of the Earth one arrives at a figure of about 4000 million years for the age of the Earth. The uranium and thorium present at the time of formation of the Earth originated in the huge stellar explosions called supernovae.

There is another important source of radiation, the so-called 'cosmic rays' from outer space. At the surface of the Earth, about 10 000 million neutrinos pass through your fingernail (about one square centimetre) per second. Fortunately, these neutrinos interact so rarely that they do not pose any health hazard! Potentially somewhat more harmful are cosmic ray muons – particles like heavy electrons – and about one muon passes through one square centimetre per minute at sea level. In fact, these muons are created at the top of the atmosphere by collisions of the primary cosmic ray particles (very high energy protons) with the molecules in the atmosphere. Almost all of the primary cosmic rays are absorbed by the atmosphere and only the relatively harmless muons reach the Earth's surface. Nevertheless, this perpetual cosmic ray bombardment is responsible for part of the radioactivity in all living things. This is because humans and all other living things contain carbon, and some of the carbon dioxide in the air we breathe contains the radioactive isotope carbon 14 (^{14}C). Since this has a half-life of 5730 years, it would have decayed away long ago if it were not constantly being replenished by cosmic ray collisions. Thus, in all of us there is a small amount of radioactive ^{14}C. When animals or plants die, no more new ^{14}C is absorbed, and from this time on the radioactive carbon just decays away without replacement. This is the basis for archaeological radiocarbon dating for time scales of several tens of thousands of years. The method was developed by Willard Libby around 1948 and produced the 'first radiocarbon revolution' since many dates were found to be much earlier than had then been thought. One problem with the method is that the rate of cosmic ray bombardment of the Earth has not been constant over these time scales. Consequently, the amount of radioactive carbon present at a given time is now determined by comparison with tree ring ages of the oldest living trees. These are the bristlecone pines found in the White Mountains of California. The combination of these two methods produced the 'second radiocarbon revolution' since some dates were now found to be even earlier than before. Radiocarbon is useful for archaeological dating back to about 35 000 years ago. For measuring ages of millions of years the decay of a radioactive isotope of potassium into argon can be used.

* *From* Physics of the Atom *by Wehr, Richards and Adair*

Fig. 5.24 A cloud chamber photograph of a cosmic ray shower. The primary particles enter at the top and develop a shower of secondary particles as they pass through a series of brass plates in the chamber.

Fig. 5.25 The bristlecone pines of the dry and inhospitable White Mountains of California are believed to be the oldest living things. Some trees are over 4000 years old.

6

Pauli and the elements

It is the fact that the electrons cannot all get on top of each other that makes tables and everything else solid.
Richard Feynman

Electron spin and Pauli's exclusion principle

Over a century ago, a Russian chemist, Dimitri Mendeleev, invented a teaching-aid for students struggling with inorganic chemistry. He realized that the properties of the 63 elements then known were repeated 'periodically' as their atomic weight increased. In other words, elements with similar chemical properties were not close together in mass, but instead were found at regular intervals as the mass increased. For example, helium is an 'inert gas' and has a nucleus with two protons. The next inert gas is neon with ten protons, then argon with 18, and so on, all with increasing mass. Mendeleev was therefore able to group all the elements into distinct families, and his scheme became known as the 'periodic table' of the elements. All good theories should be able to make predictions and this was no exception. Because of the regularities he had observed, Mendeleev realized that the then known list of elements must be incomplete. He therefore left gaps in his table corresponding to as yet undiscovered elements, and had the satisfaction of seeing gallium, scandium and germanium discovered during his lifetime. Nevertheless, the origin of the periodic table remained a mystery for over 50 years until an Austrian physicist named Wolfgang Pauli put forward his famous 'exclusion principle'. This has not only made it possible for physicists to gain an

ОПЫТЪ СИСТЕМЫ ЭЛЕМЕНТОВЪ.

ОСНОВАННОЙ НА ИХЪ АТОМНОМЪ ВѢСѢ И ХИМИЧЕСКОМЪ СХОДСТВѢ.

			Ti = 50	Zr = 90	? = 180.
			V = 51	Nb = 94	Ta = 182.
			Cr = 52	Mo = 96	W = 186.
			Mn = 55	Rh = 104,4	Pt = 197,4.
			Fe = 56	Ru = 104,4	Ir = 198
			Ni = Co = 59	Pl = 106,6	Os = 199.
H = 1			Cu = 63,4	Ag = 108	Hg = 200
	Be = 9,4	Mg = 24	Zn = 65,2	Cd = 112	
	B = 11	Al = 27,4	? = 68	Ur = 116	Au = 197?
	C = 12	Si = 28	? = 70	Sn = 118	
	N = 14	P = 31	As = 75	Sb = 122	Bi = 210?
	O = 16	S = 32	Se = 79,4	Te = 128?	
	F = 19	Cl = 35,5	Br = 80	I = 127	
Li = 7	Na = 23	K = 39	Rb = 85,4	Cs = 133	Tl = 204
		Ca = 40	Sr = 87,6	Ba = 137	Pb = 207.
		? = 45	Ce = 92		
		?Er = 56	La = 94		
		?Yt = 60	Di = 95		
		?In = 75,6	Th = 118?		

Fig. 6.1 'Experimental System of Elements' – a paper sent by Mendeleev to Russian physicists and chemists. The main difference between this and our modern 'periodic table' of elements (apart from orientation) is caused by Mendeleev's not knowing of the inert gases such as helium and neon.

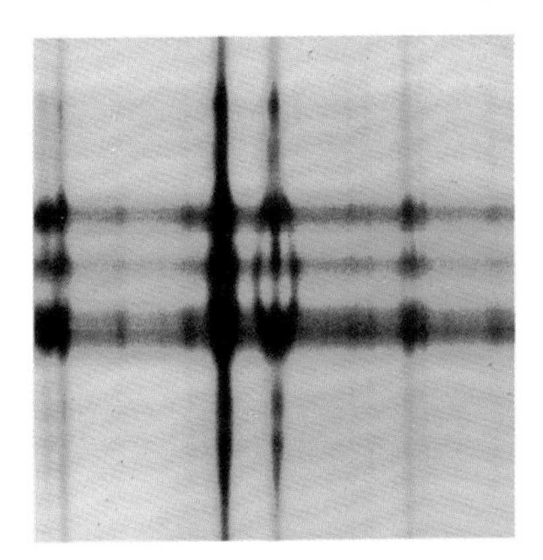

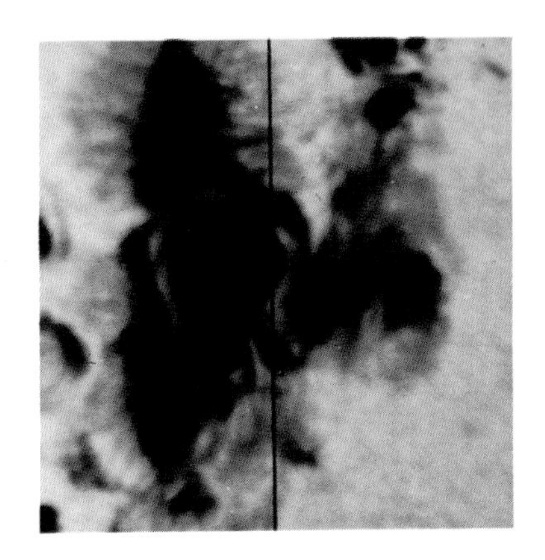

Fig. 6.2 The Zeeman effect may be used to establish the existence of magnetic fields on the Sun. When the telescope is directed at the sunspot group shown on the right, Zeeman splitting of the spectral lines is clearly visible in the associated spectrum of light from the region, shown on the left.

understanding of the different types of solids – metals, insulators and semiconductors – but also enabled nuclear physicists to explain similar 'periodicities' in the properties of nuclei. These nuclear periodicities show up as nuclei that are unusually stable against radioactive decay. These nuclei are found to contain 2, 8, 20, 28, 50, 82 or 126 protons or neutrons – the so-called 'magic numbers'. But before we look at how quantum mechanics enables us to understand Mendeleev's periodicities of the elements, we must look at another discovery that Pauli had a hand in – this time in a negative way!

In classical electricity and magnetism it is well known that a small loop of wire carrying an electric current acts like a little magnet. It was therefore possible to calculate the magnetic properties expected for electrons moving in Bohr's orbits. In 1894, long before Bohr's model of the atom, a Dutchman, Pieter Zeeman, had found that the spectral lines of atoms appear split when the atom is in a magnetic field. Zeeman's first results could be explained in terms of the different angular momentum of electrons in Bohr orbits. But there were later results – too many spectral lines – that could not be explained in this way. In true physicists' jargon this mystery was called the 'anomalous Zeeman effect'. To illustrate just how bewildering and mysterious all this was at the time, it is appropriate to tell one of the many famous 'Pauli' stories. A friend of Pauli's saw him sitting on a park bench in Copenhagen looking dejected and asked what was making him unhappy. Pauli replied 'How can one avoid despondency if one thinks of the anomalous Zeeman effect?'

Dimitri Mendeleev (1834–1907) was the youngest of a very large family of between 14 and 17 children! His fame as a chemist allowed him to escape the usual penalties for unorthodox behaviour and his liberal views and support for student causes were tolerated by the Tsarist Russian government. In 1876 he was even able to divorce his wife and marry a young art student without being pursued by the authorities. Perhaps his most endearing eccentricity was that of allowing himself only one haircut a year.

The answer to the puzzle was provided by George Uhlenbeck and Sam Goudsmit. They suggested that besides angular momentum due to its orbital motion round the nucleus, the electron also had a 'spin' angular momentum, like the spin of the Earth on its axis as it goes round the Sun. This idea was proposed in 1925, the year before Schrödinger published his wave equation. Up to that time, all the theory one had was a half-baked mixture of classical physics combined with Bohr's quantum rules. Uhlenbeck and Goudsmit gave their paper to their professor, Ehrenfest, for his comments, and he suggested they ask the great expert, Lorentz. After thinking the idea over for a week, Lorentz gently pointed out many serious difficulties for a classical picture of a rotating electron. Uhlenbeck and Goudsmit then rushed off to Ehrenfest to withdraw their paper, only to hear him say 'I have already sent your letter in long ago; you are both young enough to allow yourselves some foolishness!' In the end, of course, they were proved to have the right idea, in that all the mysterious results on the Zeeman effect are due to the electron having some extra angular momentum. However, as with Bohr orbits, a classical picture of a rotating

Fig. 6.3 Putting electrons in a box. The electrons must fill up the energy levels according to the Pauli exclusion principle. (a) One electron in the box. This can have either spin up or down. (b) Two electrons in the box. Both can go into the ground state but their spins must be oppositely directed. (c) Three electrons in a box. The ground state is full so the third electron must go into the first excited state.

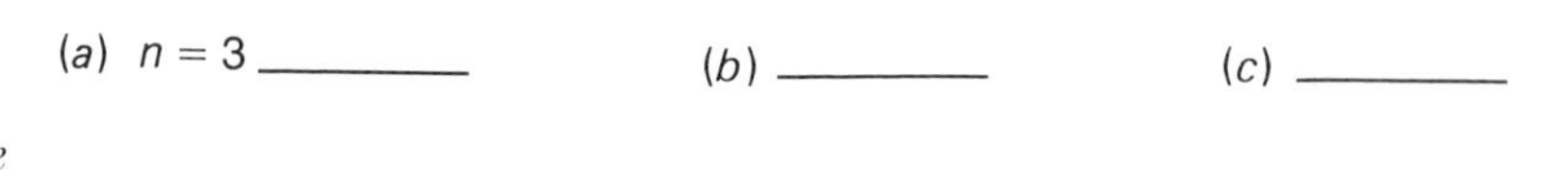

n = 2

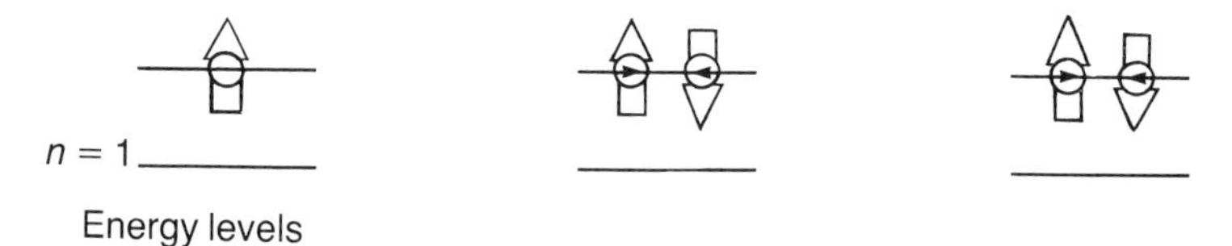

Wolfgang Pauli (1900–58) wrote a classic paper on general relativity while still a teenager. He was born in Austria, the son of a chemistry professor. The Pauli exclusion principle, proposed in 1925, explained much of chemistry and made the periodic table of the elements intelligible. Pauli obtained rather belated recognition of this fundamental contribution to quantum mechanics with the award of the Nobel Prize in 1945. He also proposed the existence of the neutrino to explain puzzling features of radioactive decays. By the time its existence was confirmed experimentally, over 20 years after Pauli made his conjecture, most physicists had already accepted its necessity. In the above photograph Pauli and his wife are shown attending the Nobel Prize ceremony in Stockholm.

Enrico Fermi (1901–54) was unique among his generation in that he did brilliant work in both experimental and theoretical physics. In his early experimental work he used the newly discovered neutrons to induce artificial radioactivity. Winning the Nobel Prize in 1938 enabled him to escape Fascist Italy and settle in the USA. As part of the war effort for the atomic bomb project, Fermi built the first nuclear reactor. The establishment of the first self-sustaining chain reaction was announced in a coded telegram sent out by Compton: 'The Italian navigator has entered the new world'.

electron cannot be taken too literally to describe quantum mechanical spin. Uhlenbeck and Goudsmit were luckier than another young physicist named Kronig who had the same idea at about the same time. He had the misfortune to ask Pauli's opinion on the matter – and Pauli convinced him that such a classical idea could not be right!

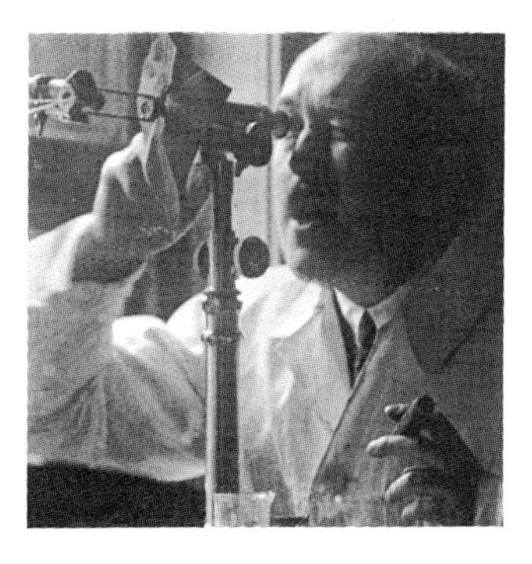

Otto Stern (1888–1969) was one of the major experimental physicists of this century. His most important work used molecular beams to demonstrate the quantum properties of atoms. In 1933 Stern was compelled to leave Nazi Germany and he moved to the USA. He won the Nobel Prize in 1943. On the right is a photograph of the postcard sent by Gerlach, Stern's collaborator in their famous experiment on 'space quantization' to Niels Bohr, announcing their discovery.

Abs. Prof. W. Gerlach
Frankfurt a. M.
Physikal. Institut
Universität

Professor
Niels Bohr
Kopenhagen
Stockholmsgade 37

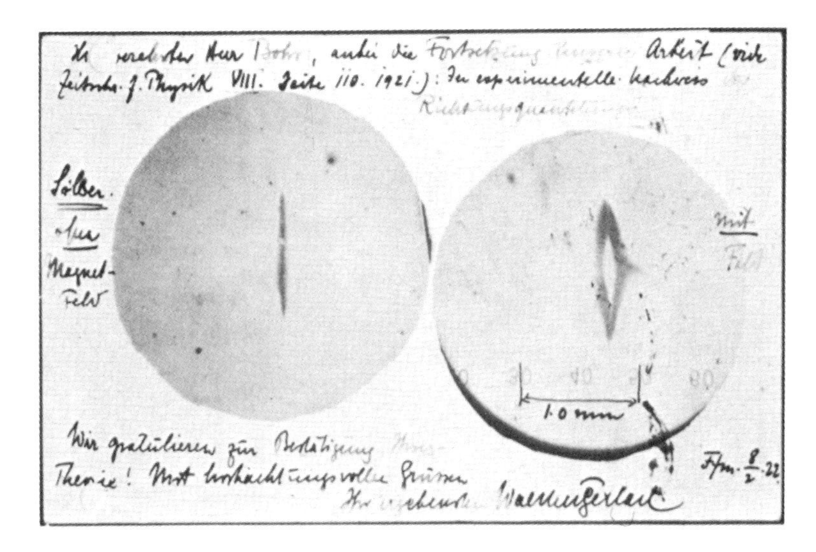

Sehr verehrter Herr Bohr, anbei die Fortsetzung unserer Arbeit (vide Zeitschr. f. Physik VIII. Seite 110. 1921.): Die experimentelle Nachweis der Richtungsquantelung.

Silber. ohne Magnet-Feld

mit Feld

1.0 mm

Wir gratulieren zur Bestätigung Ihrer Theorie! Mit hochachtungsvollen Grüssen Ihr ergebenster Walther Gerlach

Ffm. 8/2. 22

There is one more discovery about spin angular momentum that we must mention before we can get down to Pauli's exclusion principle and the periodic table. As we discussed in chapter 4, angular momentum is 'quantized' and the axis of rotation can only point in certain directions. It was Pauli who had suggested this 'space quantization' and it was verified in a famous experiment by Stern and Gerlach. For an electron, this has the consequence that there are only two possible directions of rotation: it can either spin clockwise or anticlockwise. We often talk about the electron spinning clockwise as a 'spin up' electron, and anticlockwise as a 'spin down' electron. This property of the electron provided Pauli with the final clue to understanding the structure of atoms.

The essential problem to be explained had been pin-pointed by Niels Bohr. If the energies of electrons in atoms are indeed quantized, why is the lowest energy state of an atom not one with all the electrons in the lowest energy level? It was obvious that all the electrons were not in the lowest state, since, if they were, all the elements would behave in a similar way. Moreover, as we shall see, it is the shape of the excited state wavefunctions that enables atoms to combine to form molecules. If all electrons were in the symmetrical,

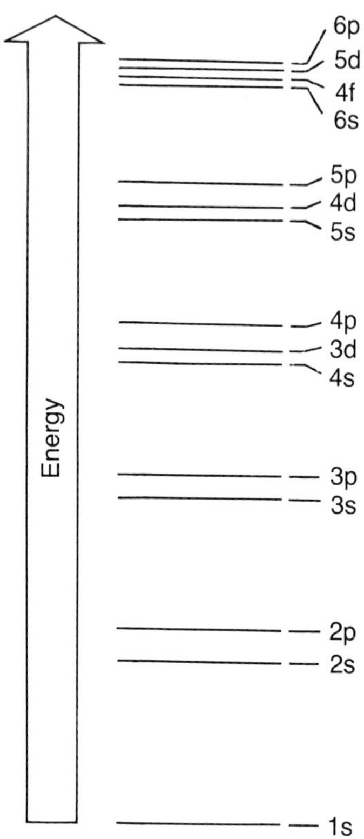

Fig. 6.4 Energy levels for a typical atom. This diagram, along with the Pauli principle, determines the form of the periodic table of the elements.

lowest energy state there would be no molecules and certainly no life as we know it! Pauli provided the answer with his exclusion principle. This is the assertion that only one electron is allowed in each quantum state. Consider what this means for electrons in a box – the quantized energy levels were discussed in chapter 4. With one electron in the box, the lowest energy (ground state) of the system has this electron in the $n=1$ level, with its spin either up or down. When adding a second electron to the box, we must obey Pauli's exclusion principle. This can also go in the $n=1$ state, provided its spin is opposite to that of the first electron. However, when a third electron is put in the box it cannot go into the $n=1$ level since this is now full. To put it there with either spin up or down would mean that two electrons had exactly the same quantum numbers, and this is forbidden by Pauli. It must therefore do the next best thing and go into the lowest available empty energy level – in this case one of the two possible $n=2$ level spin states. And so on. As Feynman says at the beginning of this chapter, it is the Pauli exclusion principle that makes all things hard and rigid. In essence, all it says is that 'matter' cannot be compressed to nothing but must occupy a certain minimum space. All 'matter-like' quantum particles will therefore obey the exclusion principle – such particles are called 'fermions' in honour of Enrico Fermi, who was one of the first to look at the implications of Pauli's principle. In fact, there is another class of 'particles' which we can categorize as 'radiation-like' (photons are an example) to which the Pauli principle does not apply. Such particles are known as 'bosons', after an Indian physicist, Satyendra Bose. In contrast to fermions, bosons prefer to be all together in the lowest energy state if at all possible!

The elements

We are now in a position to understand not only the variety of elements found in Nature but also their chemical properties. Any detailed understanding of the periodic table requires some knowledge of wavefunctions and quantum numbers. These were discussed in the final section of chapter 4, and, like that section, this one may be rather heavy going at a first attempt. As before, it is probably best to skim quickly through this section and not worry about the details. Only a few references to the results of this section are made in later parts of this book. This section attempts to explain how the Pauli exclusion principle can account for different types of chemical bonding and how the electrons fill up the available empty energy levels to yield the different elements. For hydrogen, we have seen in chapter 4 how the Schrödinger equation led to Bohr's quantized energy levels. Now for a nucleus with Z protons we will need to add Z electrons to make a neutral atom and, according to Pauli, these will not all go into the lowest energy state. Instead, they will fill up the energy levels starting from the lowest $n=1$ level, allowing a spin up and a spin down electron in each state labelled by the quantum number n and the two orbital angular momentum quantum numbers L and M. In actual fact, the energy level diagram that will dictate the order of level filling for a many-electron atom will not look quite the same as that for hydrogen. This is because, in addition to the attractive force on an electron from the nucleus, each electron will feel repulsive forces from all the other negatively charged electrons. One effect of this is that an electron in one of the higher n levels – corresponding to a large Bohr orbit – only sees a fraction of the nuclear charge. The positive charge of the nucleus is 'screened' by the negative charges of other electrons closer in. Moreover, the $L=0$, or 'S state' electrons, have probability distributions that are greater nearer the nucleus than any $L=1$ ('P state') or $L=2$ ('D state') electrons (see fig. 4.15). The S electrons will therefore 'see' more of the nuclear charge and be more tightly bound. Thus, we expect the energy level pattern for many-electron atoms to look something like fig. 6.4. All we have to do now to

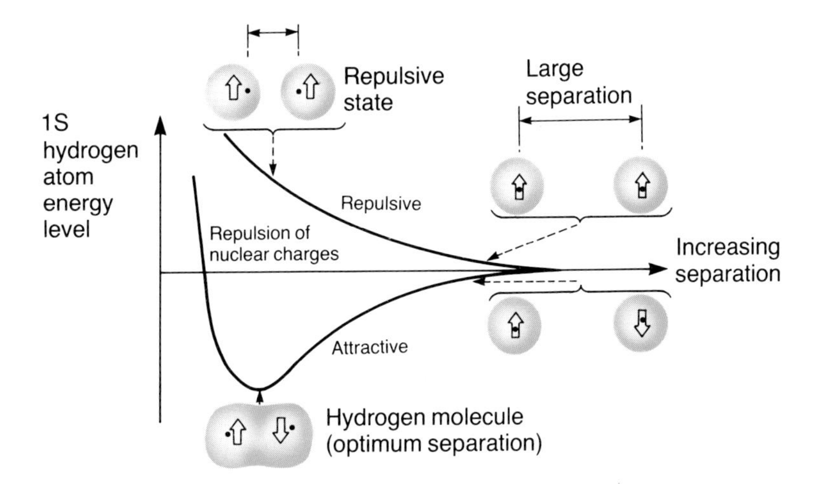

Fig. 6.5 Potential energy curve for two hydrogen atoms versus their separation. The atoms repel each other when the two spins are parallel and attract when they point in opposite directions. The 1S level of the hydrogen atom therefore splits into two as shown. If the atoms are brought too close together the proton nuclei will start to repel each other significantly. Thus, in the case when the two spins are opposite, there is an optimum separation for formation of a hydrogen molecule.

explain the periodic table is to fill up these levels according to Pauli's housing plan. It was not for nothing that Pauli had the nickname 'atomic housing officer'!

A neutral hydrogen atom has one electron which is normally found in the $n=1$ and $L=0$ 'ground state': the lowest '1S' energy level. The electron can be 'excited' to one of the higher energy levels by collisions, or by shining light on to the atom. After a very short time, the electron will drop back into the ground state, emitting a photon with an energy corresponding to one of the spectral lines, as we discussed before. Perhaps surprisingly, however, the Pauli principle has a role to play even for hydrogen. What happens if we bring up another hydrogen atom close to the first one? If the two electrons are both in the spin up state, the exclusion principle must prevent the two atoms getting close enough for the wavefunctions of the two electrons to overlap, since this would mean that the two electrons would be in the same quantum state. Conversely, if the two spins are opposite, a close approach is possible, and indeed the two electrons spend most of their time in the region between the two hydrogen nuclei. This results in a binding force between the two hydrogen atoms and allows a stable hydrogen 'molecule' to be formed. This type of chemical bond – in which the two electrons are 'shared' between the two nuclei of the molecule – is called a 'covalent' bond. It is Pauli's exclusion principle that explains why hydrogen is chemically active and why two atoms can form a stable H_2 molecule. Notice that the same principle also forbids a third atom to form a covalent bond with the H_2 molecule since both the two ground state spin states are already occupied.

The next simplest element is helium, with two electrons surrounding its nucleus, which has a positive charge twice that of the hydrogen nucleus. These two electrons can both go in the lowest 1S level provided their spins are opposite. Since there is now no more room for any other electrons in the 1S state, the Pauli principle will tend to keep other electrons away from the helium atom, just as it does for the H_2 molecule. Thus, we expect helium to be chemically inactive – one of the family of 'inert gases'. The next element, lithium, has three electrons surrounding its nucleus and has an energy level structure like that also shown in fig. 6.6. The first two electrons can both go into the 1S state, with opposite spins, and this now forms a chemically inactive 'closed shell', like that of helium. The third electron must go into the lowest unoccupied energy level, which is now the 2S level. Thus, lithium has one electron in an $L=0$, S level and this explains why it has similar chemical properties to hydrogen. For example, lithium forms stable Li_2 molecules through covalent bonding in the same way as hydrogen combines to form H_2 molecules.

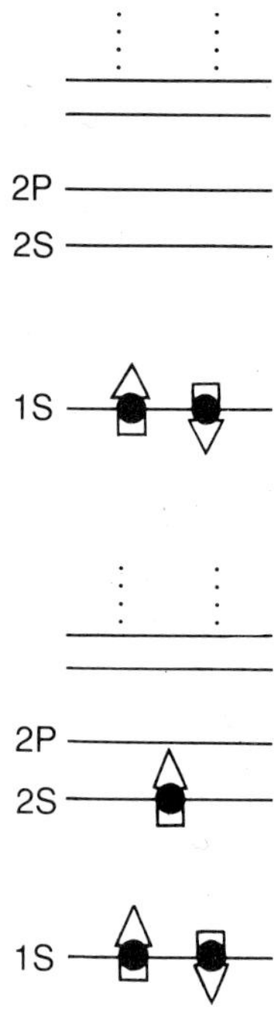

Fig. 6.6 Helium and lithium energy levels.

If we keep adding successive electrons, we begin to fill higher and higher energy levels. Nitrogen, for example, has seven electrons. Two of these electrons fill up the 1S level to form a closed shell. Two more fill up the 2S level to form another closed shell, leaving three electrons to be put into the 2P level. Now, S states have electron probability distributions that are 'spherically symmetrical' – no direction is favoured (see fig. 4.15). The 2S state is larger than the 1S state, corresponding to the fact that the excited state is less tightly bound than the ground state. However, the P state probability patterns are not spherically symmetrical. As we saw in chapter 4, there are three possible P states, which we can specify by labels x, y or z. These labels tell us whether the 'lobes' of the electron probability pattern are oriented along the x, y or z directions (see fig. 4.16). In order to get the three electrons as far apart as possible, and thereby minimize the electron repulsion between them, the three electrons in nitrogen go into three different P states, in preference to two of them going into the same P state with their spins opposite. The shapes of these P state wavefunctions also allow us to understand the formation of more complicated molecules. A hydrogen atom can approach a nitrogen atom and attach itself to any of the three P state lobes, provided its spin is opposite to that of the corresponding P electron. Indeed it is clear that nitrogen can attach a maximum of three hydrogen atoms to itself before the P shell has a full complement of six shared electrons. Fig. 6.7 shows the geometry of the ammonia molecule, NH_3,

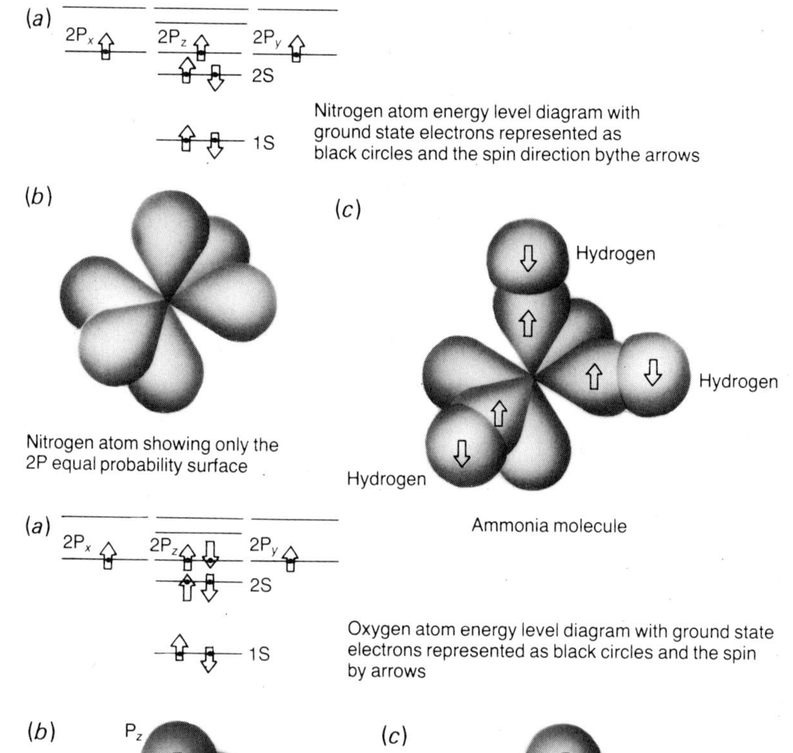

Fig. 6.7 An attempt to show the structure of the ammonia molecule NH_3. (a) Energy levels for nitrogen. (b) Sketch of 2P probability surfaces for a nitrogen atom. (c) Arrangement of electron probability surfaces in an ammonia molecule.

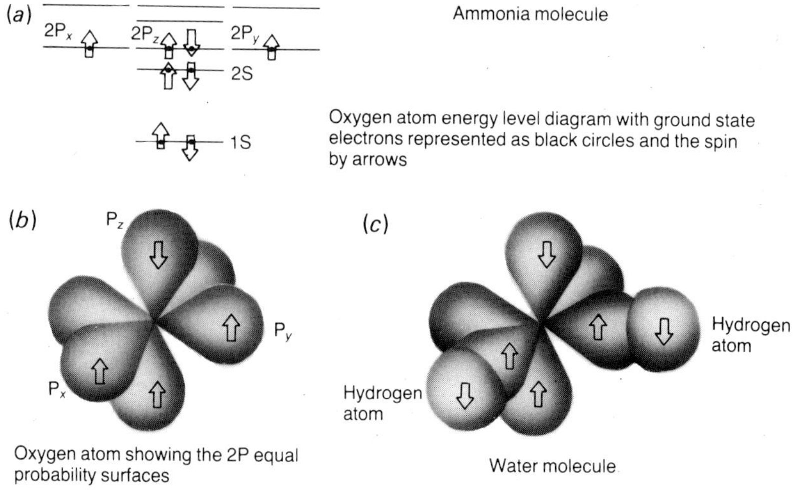

Fig. 6.8 Attempt to illustrate the structure of a water molecule. (a) Energy levels for oxygen. (b) Sketch of 2P probability surfaces for an oxygen atom. (c) Arrangement of electron probability surfaces in a water molecule.

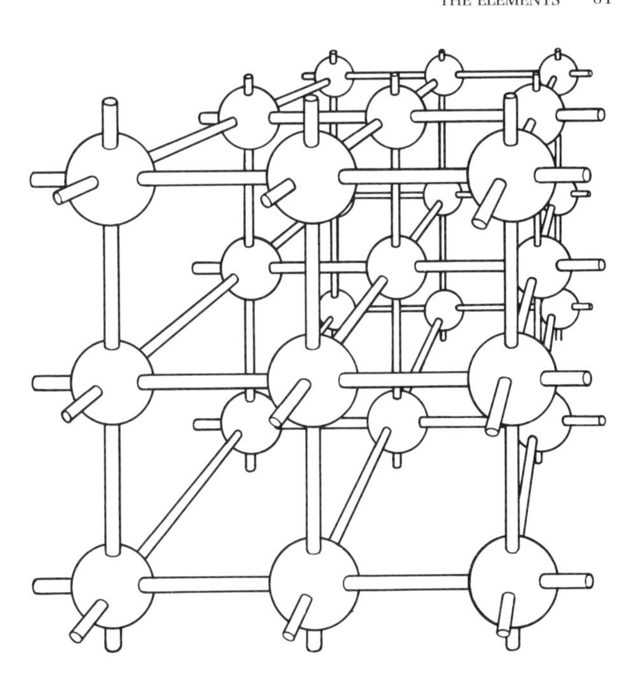

Fig. 6.9 The arrangement of atoms in a cubic crystal lattice. The spheres represent the location of the atoms and the tubes linking them represent the direction of the bond lines holding the atoms in place.

Fig. 6.10 Common table salt magnified 50 times by an electron microscope. The evident cubic structures reflect the underlying lattice structure.

produced by the covalent bonding of one nitrogen atom and three hydrogen atoms. Clearly, nitrogen will be chemically active and able to form many other compounds. After nitrogen, adding another electron brings us to oxygen with four P shell electrons. One of the three P 'orbitals' must now be full, and consequently the oxygen atom can only attach itself to two hydrogen atoms. An impression of such a water molecule, H_2O, formed in this way is shown in fig. 6.8.

We now move on through the periodic table to neon; this has ten electrons to put into energy levels according to Pauli. The 1S, 2S and 2P shells are now completely filled. We therefore understand why neon, like helium, is chemically inactive, and why the property of 'inertness' has been repeated. From the energy level pattern shown in fig. 6.4 we see that another inert element should occur when we have filled up both the 3S and 3P levels. With two electrons in an S shell and six in a P shell this brings us to an element with proton number $Z=18$: the inert gas argon. The whole of the periodic table can be understood in this way. Elements have similar chemical properties when they have the same number of 'outer' electrons in similar quantum states. Thus, lithium combines with oxygen to form 'lithium oxide', Li_2O, in exactly the same way that hydrogen combines with oxygen to form a water molecule.

Metals, insulators and semiconductors

One of the great successes of quantum physics has been in understanding the way in which different types of solids conduct electricity. In a solid it is the flow of electrons that gives rise to electric currents. It is a great triumph of quantum mechanics that it can explain what makes different substances metals, insulators or semiconductors. Indeed, it is no exaggeration to say that it is this quantum mechanical understanding that has led directly to the

Fig. 6.11 This Landsat photograph shows San Fransisco Bay with Silicon Valley and San Jose at the top right. The San Andreas Lakes that lie along the infamous earthquake fault are clearly visible slanting upwards parallel to the coast in the bottom right. The SLAC accelerator can also be picked out as a line crossing the freeway between San Jose and San Fransisco. The Golden Gate Bridge is also visible.

present technological revolution, with its accompanying avalanche of new and cheap 'solid-state' devices ranging from stereos and colour TVs to all sizes of computers. There are many properties of solids that can be understood from a quantum mechanical viewpoint – colour, hardness, texture and so on – but we are going to focus on their ability to conduct electric currents. A good conductor, such as copper, must have many 'conduction' electrons that are able to move and carry a current when an electrical potential difference is applied to a wire. An insulator such as glass or polythene, on the other hand, apparently has no conduction electrons at all, since no current flows when a voltage is applied. There is a third category of materials which consists of solids that conduct electricity much better than insulators but much worse than metals. Not unnaturally, such materials are called semiconductors. Germanium and silicon are examples of semiconductors, and their importance to much of our new technology is evident from the renaming of the area around San Jose in California as 'Silicon Valley'.

The properties of a solid depend not only on what it is made of but also on the way the atoms or molecules are stacked together. Many materials have their constituent atoms arranged in a regular array – like the pattern of bricks in a wall. This regular pattern of atoms is called a 'crystal lattice', and substances with such a structure are called 'crystalline solids'. There are other substances that have no crystalline structure, but, just like a pile of bricks, still have a certain strength and hardness. Because of their lack of an underlying crystal structure the properties of these 'amorphous' solids are much more varied than those of the crystalline solids we shall talk about in this chapter. As we shall see, arranging all the atoms in a regular array has a dramatic effect on the allowed energy levels for the atomic electrons.

Fig. 6.12 The beautiful symmetry of this snowflake reflects the hexagonal bonding structure of the water molecules making up the ice.

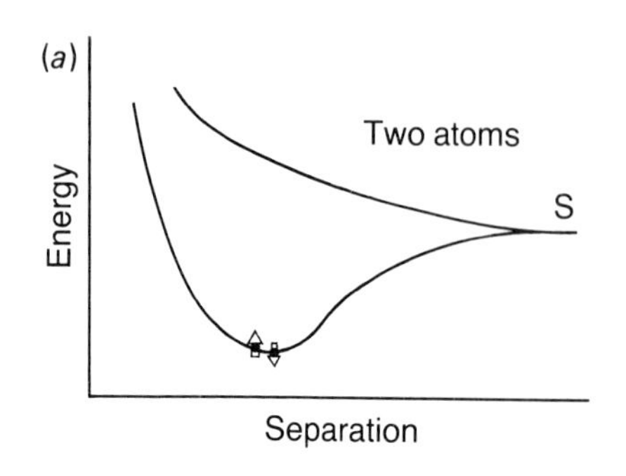

S level splitting as two atoms approach each other

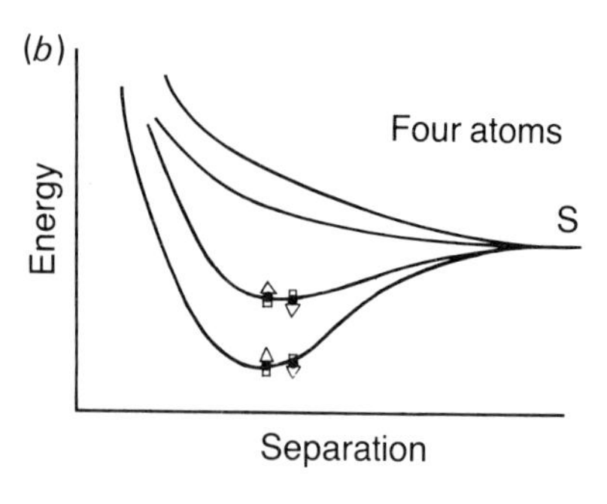

S level splitting as four atoms approach each other

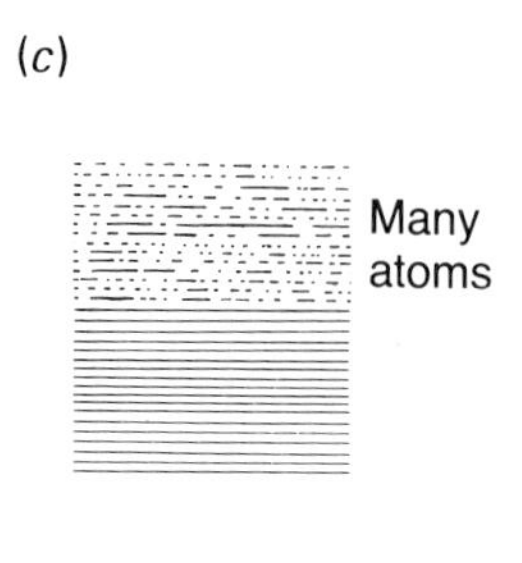

Formation of a half-filled conduction band for a metal like sodium

Fig. 6.13 Level splitting for an element such as sodium with a single S state electron in its outer shell. (a) *Level splitting as two atoms approach each other.* (b) *Splitting of levels due to four atoms approaching each other.* (c) *Formation of a half-filled 'conduction band' as many atoms approach each other.*

We can get an idea of the energy level structure of a collection of atoms by looking at what happens to the atomic energy levels of two atoms as they are brought together. In the case of hydrogen, we saw that the Pauli exclusion principle leads to binding of the two atoms to form a molecule only when the spins of the two electrons are opposite. If the spins are parallel, the Pauli principle keeps the electrons apart and there is no binding. In terms of energy levels, we see that, in one case, the two electrons have an energy less than that of two isolated atoms – resulting in covalent binding to form a molecule – and, in the other case, the electrons have more energy than two atoms – hence giving no binding (see fig. 6.5). A similar effect takes place for the energy level of the outermost 3S electron of sodium when we bring together two sodium atoms. Moreover, when we bring up more and more sodium atoms, the 3S levels split up more and more until all that remains is a 'band' of very closely spaced levels. This band is called the '3S band' since these levels were formed from the 3S levels of atomic sodium. For N atoms, the 3S band will contain N levels, each of which can hold two electrons, spin up and spin down. The lower lying atomic energy levels correspond to more tightly bound electrons with smaller wavefunctions which do not overlap as much as those of the 3S states. The resulting bands are much narrower. The 1S and 2S bands can accommodate $2N$ electrons and, for sodium, will be fully occupied. The 2P band can hold $6N$ electrons (three different P states times two different spin states for each of the N atoms) and will also be full. However, a sodium atom has only one electron in the 3S state so that, in a metal of N atoms, the 3S band will contain N electrons and will be only half full. These 3S electrons are the 'conduction' electrons. If we apply a potential difference to a wire made of sodium metal, these electrons can gain energy and be accelerated in the direction of the potential – we can imagine them jumping to the empty higher energy levels in the 3S band. This picture for the energy levels of conduction electrons in a metal is the reason why many properties can be explained by the simple model of electrons in a 'box' that we considered in an earlier chapter. In the covalent binding for the hydrogen molecule, the two electrons are shared between the two hydrogen atoms. In a sense, therefore, metals can be thought of as an extreme case of covalent binding in which the conduction electrons are shared by all the atoms in the metal.

The situation we have described above corresponds to the lowest energy state of sodium metal. At room temperatures, however, the lattice ions have some thermal kinetic energy corresponding to 'vibrations' of the ions about their positions in the crystal lattice. The conduction electrons are therefore

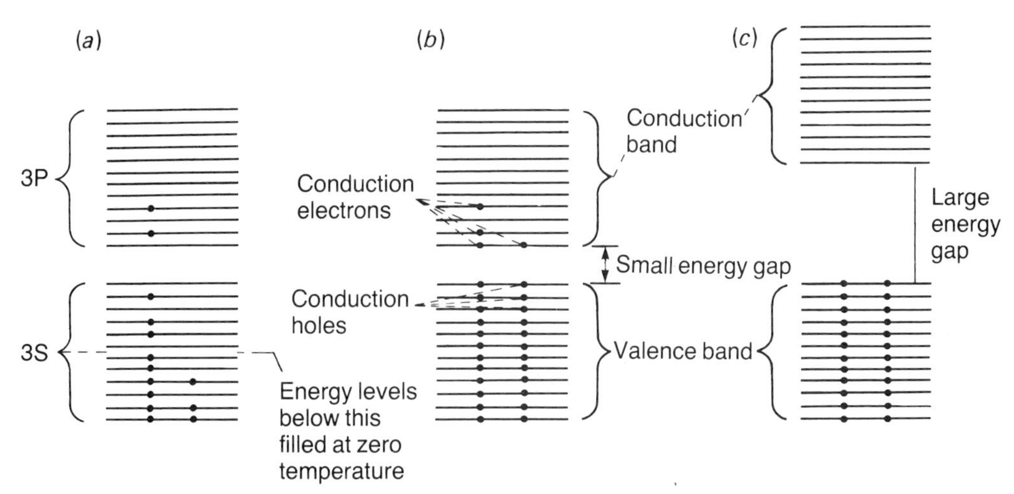

Fig. 6.14 Metals, semiconductors and insulators. (a) *Band structure of a typical metal like sodium. There are many unfilled energy levels for the electrons to move into. At normal temperatures a few of the electrons will be excited into the almost unoccupied 3P level.* (b) *In a semiconductor the valence band is full and there is a small energy gap to the empty energy levels in the conduction band. At normal temperatures some of the electrons have enough energy to jump across the gap.* (c) *In an insulator the gap between bands is too large for there to be a significant number of electrons able to jump the gap. It therefore conducts electricity very poorly if at all.*

able to gain and lose energy in collisions with the lattice ions and with each other. Thus, instead of the conduction electrons filling up exactly the bottom half of the 3S band energy levels, with the top half being empty, some of the electrons will be 'thermally excited' to these higher levels. This also, of course, has the effect of leaving some empty levels in the bottom half of the band. Although the energy involved in a typical collision at room temperature is only a fraction of an electron-volt, the gaps between the bands in sodium are nevertheless small enough for some of the electrons to be excited to the previously empty 3P band from the 3S band.

This complication of thermal excitation of electrons does not change our picture of conduction electrons in metals significantly, but it will be crucial for an understanding of insulators and semiconductors. Let us first see how this simple energy band picture, together with the Pauli exclusion principle, can give us an explanation of insulators. Consider what happens if we have a material in which the ground state consists of one band being completely full and the one above completely empty. If there is a large energy gap between the bands, almost no electrons are able to gain enough energy from collisions to jump up into the empty band. Thus, when a voltage is applied to the material, there are no empty energy levels for the electrons to jump into and so gain energy – since Pauli does not allow two electrons to occupy the same quantum state. The lower band is already full and there is too large an energy for an electron to jump up to the higher empty band. This is the situation in an insulator: there are essentially no free conduction electrons to contribute to an electric current in the upper 'conduction band'. What is a semiconductor? It is a material with a similar band structure to an insulator, but where the energy gap between the bands is much smaller. At ordinary temperatures, therefore, a significant number of electrons are excited to the upper conduction band. Now when a voltage is applied, there are electrons in the upper band with plenty of empty states to move to and allow the electrons to gain energy. There are also empty states in the lower band which allow conduction in this band as well. Thus, semiconductors will conduct currents fairly easily and their conductivity will depend strongly on temperature, in contrast to metals and insulators.

The picture of bands given above seems to be based strongly on the atomic energy levels of the material. One therefore expects metals to have odd numbers of electrons, while insulators and semiconductors should correspond to elements with closed shells. In fact, magnesium, with a filled 3S shell, is a good conductor, while carbon, with only two electrons in its 2P shell, is an insulator! The answer to these puzzles lies in the details of how

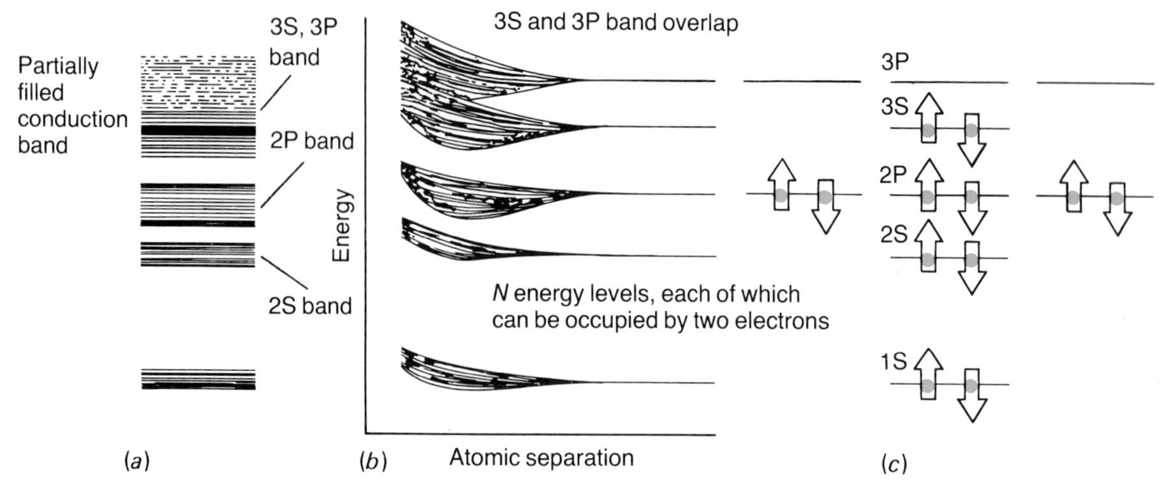

Fig. 6.15 Band overlap in magnesium. (a) Energy levels for isolated magnesium atom. (b) Variation of the energy level diagram for N magnesium atoms with their separation. (c) Band structure for magnesium.

some bands can overlap and leave no energy gap. In magnesium, it is the 3S and 3P bands that overlap to yield a single band capable of holding $2N + 6N = 8N$ electrons. Since only $2N$ of these levels are filled, magnesium is a good conductor. For carbon, on the other hand, as the N atoms are brought together, the 2S and 2P bands join to form a band with $8N$ states, as for magnesium. However, as the carbon atoms come still closer together, this combined band splits into two bands each with $4N$ states. In carbon, the lower band is full and the upper band completely empty, characteristic of an insulator. The energy levels in germanium and silicon behave similarly, but here the two bands are separated by a much smaller energy gap so that both are semiconductors rather than insulators. The way to understand the detailed band structure of materials was discovered by a Swiss physicist named Felix Bloch. He solved the Schrödinger equation for electrons moving in a potential corresponding to a regular lattice of positive ions. This leads to the energy band structures described above and is the mathematical basis for the quantum mechanical 'band theory' of solids.

Transistors and microelectronics

Pure semiconductors are not in themselves of great practical importance. Only one atom in about a thousand million contributes an electron to conduct electricity: in metals almost every atom contributes one or more electrons. But this apparent drawback has the great advantage that the conduction properties of semiconductors can be 'tailored' at will by the introduction of suitable 'impurity' atoms at the level of one in a million. Germanium and silicon both have four 'valence' electrons which fill up most of the $4N$ states of the 'valence band', which lies below the almost empty conduction band. If we introduce an impurity such as phosphorous – which has five valence electrons – then, since only four electrons are needed for the covalent bonding of the lattice, there is an electron 'left over' that can easily be detached and contribute to the conductivity. Similarly, if we introduce impurity atoms with only three valence electrons, such as boron, there is now one covalent bond lacking in the lattice, and this tends to capture electrons from the valence band, leaving an empty state which will allow some conductivity. These two situations are represented on the energy level diagram shown in the figure. The phosphorus atoms give rise to 'donor states' just below the conduction band, and these electrons need only gain a small amount of energy to jump into the conduction band. Semiconductors that have been 'doped' in this way are called 'n-type' semiconductors, since

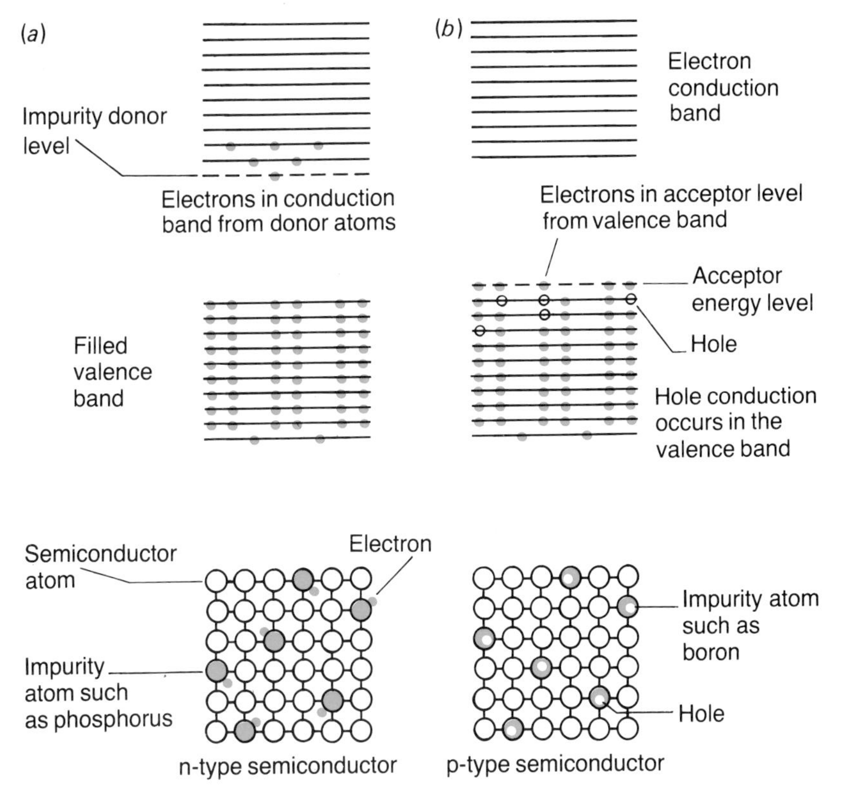

Fig. 6.16 Semiconductors doped with impurity atoms. (a) n-type semiconductor in which the impurity atoms have an extra electron. This results in the effective energy level diagram shown above the figure. (b) p-type semiconductor doped with impurity atoms with one fewer electron, resulting in electron 'holes'. The energy level diagram corresponding to this situation is shown above the figure.

they give rise to an extra contribution to the current from the negatively charged electrons from the donor levels. Semiconductors doped with boron are called 'p-type' semiconductors. The boron atoms give rise to 'acceptor states' just above the nearly full valence band, and at room temperatures electrons are readily excited into these levels. Why are they called p-type semiconductors? Well, as we said before, conductivity in the nearly full valence band is only possible because electrons can move into the few unoccupied states. As can be seen from fig. 6.16, instead of the electron moving, we can equally well think of the 'hole' moving in the opposite direction to the electron. Since moving a negative charge to the left has the effect of increasing the charge on the right, we can think of the current not in terms of negatively charged electrons moving to the left, but rather as positively charged holes moving to the right. In a p-type semiconductor, extra holes have been created in the valence band and we can think of the increased current as being due to the positively charges holes we have added to the material.

Why is all this useful? The reason is that p- and n-type semiconductors can be put together to form a 'switch' to control current flow. The simplest semiconductor device is a p–n junction diode that allows current to pass in only one direction when a voltage is applied. The p–n junction is therefore able to 'rectify' an alternating voltage. But the development which has affected people's lives most directly is the invention of the transistor by John Bardeen, Walter Brattain and William Shockley at the Bell Telephone Research Laboratories in the USA. The transistor was not discovered by accident, but instead was the culmination of an extensive research programme. As Bardeen described it in his Nobel Prize lecture, 'the general aim of the program was to obtain as complete an understanding as possible of semiconductor phenomena, not in empirical terms, but on the basis of atomic theory'. It seems a long way from de Broglie's probability waves to

The three inventors of the transistor at about the time they made their invention. From left to right, they are Shockley, Brattain and Bardeen. Bardeen went on to win a second Nobel Prize for his work on superconductivity. No one else has won two physics Nobel Prizes.

modern computers! A replica of the first 'point-contact' transistor they produced is shown in fig. 6.18*a*. This was in 1947 and it was soon followed in 1951 by the not very glamorous looking, but more reliable, 'pnp' transistor (see fig. 6.18*b*). This consists of a thin layer of n-type semiconductor sandwiched between two thicker regions of p-type material. 'Transistor action' is the turning on and controlling of a large current through the high resistance 'collector–base' p–n junction, by a small current through the low resistance 'base–emitter' n–p junction. The word transistor refers to this effect and comes from combining the two words '*trans*fer-res*istor*'.

Transistors turned out to be ideally suited, for the 'on–off' binary logic of computers. Moreover, their reliability and low power consumption, together with a number of incredible engineering advances, have now made them the basic ingredient of modern 'microelectronics'. The key idea seems to have been first written down by a British engineer named G. W. A. Dummer, working at the Royal Radar Research Establishment in Malvern, Worcestershire. He was an expert on reliability problems of electronic components and was concerned with the performance of radar equipment under extreme conditions. Dummer eventually realized that it was unnecessary to manufacture all the ingredients for an electronic circuit – transistors, resistors (which impede the flow of current) and capacitors (devices for storing charge) – in separate pieces; the circuit could be made much smaller and more robust if all these devices were contained in the same piece of semiconductor! In May, 1952, Dummer wrote 'With the advent of the transistor and the work in semiconductors generally, it seems now possible to envisage electronic equipment in a solid block with no connecting wires.

The block may consist of layers of insulating, conducting, rectifying and amplifying materials, the electrical functions being connected directly by cutting out areas of the various layers'. This is an amazingly accurate vision of a modern 'integrated circuit'. In 1952, however, there were still many difficult technical problems to be overcome before Dummer's idea could become a reality. Unfortunately, although Dummer produced a non-working model of such a silicon 'solid circuit' in 1957, the potential of such developments was not appreciated by many people in the UK. Thus, it was that the vital breakthrough in this direction was made in the summer of 1959 by an American, Jack Kilby, working for Texas instruments. Kilby created the first working 'integrated circuit' (IC). Since ICs are physically made out of tiny chips of silicon, they are popularly known as 'chips' by the trade, or as 'microchips' by the newspapers. The full utility of the IC only became apparent after the invention of a new process for making 'planar' transistors. The planar transistor was discovered late in 1958 by a Swiss-born physicist named Jean Hoerni, one of the founder members of Fairchild Semiconductor. With this invention, Robert Noyce, a co-founder of Fairchild, was able to design and produce a truly robust IC which could be mass produced. Using these ICs, Fairchild was able to market a whole family of 'logic' chips – the decision-making units of computers. That year, 1962, marked the beginning of mass production of ICs. The same year marked another technological breakthrough with the discovery of a new and more convenient form of transistor for incorporation in mass produced chips. This was the 'MOSFET', the metal-oxide–semiconductor field effect transistor, invented by two young engineers, Steven Hofstein and Frederic Heiman, at RCA's research laboratory in New Jersey. Meanwhile, chip development continued apace, with ever increasing miniaturization and complexity; by 1967 chips were being produced that incorporated thousands of transistors.

The various stages of computer development may be crudely categorized in terms of 'generations'. The first generation began in the 1950s with the first industrial computer, UNIVAC 1, constructed using electronic valves. The first IBM computer, the IBM 701, was delivered in 1953: by 1956 IBM had already become the largest and most profitable computer manufacturer, building machines by the hundreds! The widespread availability of transistors to replace the costly and unreliable valves saw the beginning of the second generation of computers in about 1959. Of course, along with these 'hardware' developments came 'software' improvements in the art of

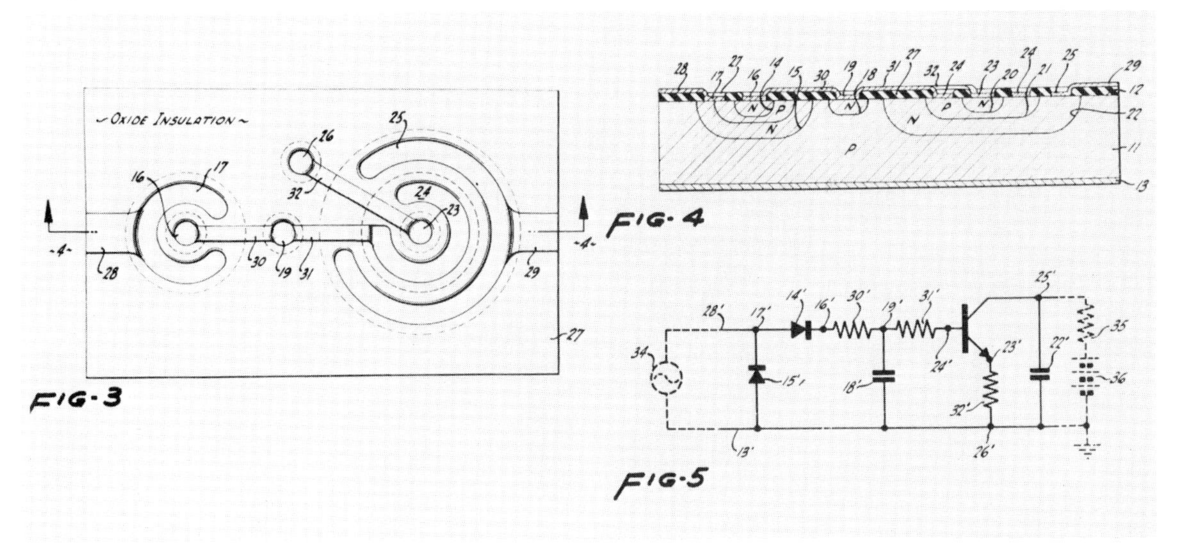

Fig. 6.17 Diagrams from Noyce's patent for integrated circuits made using a planar process. This was a crucial breakthrough in chip manufacture.

(*a*)

(*b*)

Fig. 6.18 The first transistors. (a) A replica of the point-contact transistor invented by Bardeen and Brattain. The wedge of semiconductor which forms the base is about 3 cm each side. (b) Shockley's junction transistor looks very unglamorous but was much easier to fabricate reliably.

Fig. 6.19 The first 'integrated circuit' or chip. Instead of making the components of a circuit separately, Jack Kilby incorporated a transistor, a capacitor and resistances in the same piece of germanium.

'programming' the computers – basically, how to get the computer to do what you want it to do! In about 1966, following hard on the heels of second-generation computers, came the third generation, whose main hardware innovation was the incorporation of ICs. This development made third-generation computers smaller, cheaper and far more reliable than previous generations. The most sophisticated ICs now had tens of thousands of transistors; this level of complexity is known as large scale integration or LSI for short. What could come next? Probably the best way to characterize the difference between third- and fourth-generation computers is by the invention of the 'microprocessor'. By 1968 Robert Noyce had left Fairchild to found Intel (short for *int*egrated *el*ectronics). One of Intel's employees, Ted

Fig. 6.20 The first microprocessor. Ted Hoff, an engineer with Intel, had the idea of putting all the elements of a programmable computer on one chip. The chip has dimensions of about 3 by 4 mm and it contains over 2000 transistors.

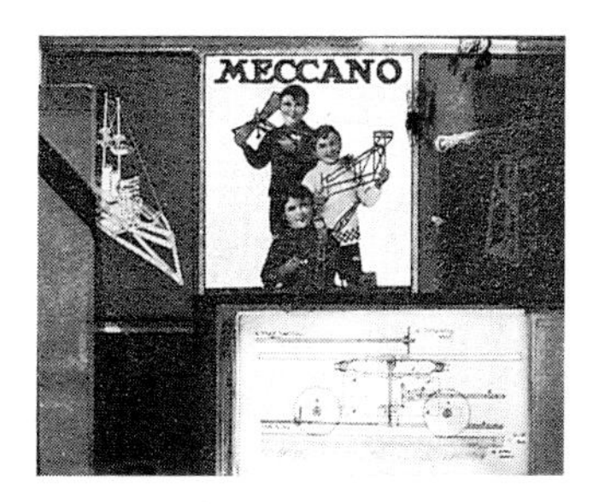

The British physicist Douglas Hartree was the first to use an automatic calculator for problems in atomic theory. During the summer of 1933 he went to MIT in the USA and used Vannevar Bush's mechanical calculator to compute 'approximate wavefunctions for mercury'. On returning to the UK, Hartree and Arthur Porter built a similar machine using £20 worth of Meccano! Hartree commented that his 'first impression of Dr Bush's machine was that . . . someone had been enjoying himself with an extra large Meccano set'. Such analogue machines were superseded by more accurate digital computers.

Hoff, Jr, had the bright idea to design a programmable IC chip. Instead of having to design a special chip to perform a specific function, a microprocessor chip can be programmed for the particular application. It took Intel some time to realize that they were sitting on a goldmine. Initially only thought of as a device for calculators and minicomputers, the microprocessor now has applications in all manner of places: washing machines, typewriters, thermostats, video games and personal computers to name but a few. The first microprocessor was marketed in 1971 and contained 2000 or so transistors; there are now microprocessor chips with as many as 500 000 transistors. Such a high level of miniaturization is abbreviated to VLSI, standing for very large scale integration, and fourth-generation computers using VLSI chips are still being developed.

Robert Oppenheimer and John von Neumann at the official dedication of the computer built for the Princeton Institute for Advanced Study in 1952. Oppenheimer was head of the Manhattan project at Los Alamos during the Second World War. Von Neumann was a brilliant Hungarian mathematician who moved to the USA before the war. He used the ENIAC computer to perform vital calculations for the design of the implosive lens of the first atomic bombs. He also set down the first specification for a stored-program computer which proved to be of major importance in the development of powerful modern computers. During the McCarthy era Oppenheimer was investigated as a security risk, primarily because he opposed the development of the hydrogen bomb. During the hearings, von Neumann testified to Oppenheimer's loyalty and integrity.

To end this chapter, let's take a final look into the future. According to Carver Mead, California Institute of Technology's guru on VLSI, there are no inherent engineering or physics obstacles to stop us fabricating chips containing millions of transistors by the end of the 1980s. With such complex chips there is the possibility of constructing enormously powerful computers. It is the goal of the ambitious Japanese 'fifth-generation' project to produce computers over a 1000 times faster than today's fastest supercomputers. Furthermore, the Japanese hope to use this vast processing power to construct machines capable of scanning rapidly through huge data and 'knowledge' stores, and able to make logical decisions based on the data they are presented with. Whether or not these 'intelligent' computers come to pass, it is certain that such computers will need to use the computing power of many computers working together. Computer scientists will need to learn how to organize such 'multi-computer' machines so that all their potential power can be used efficiently. One idea of how this might be done is based on a device called a 'transputer' (from *trans*istor and com*puter*) that is manufactured by INMOS. The transputer is a computer on a chip, together with four 'links' that enable it to be easily connected to other transputers. Using a novel programming language called *occam*, able to distribute a problem over many transputers, it may be possible to use this VLSI chip as a 'Lego' building brick for genuine fifth-generation machines.

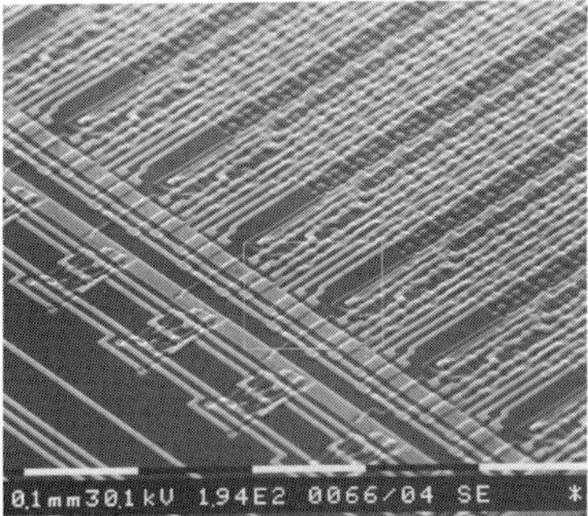

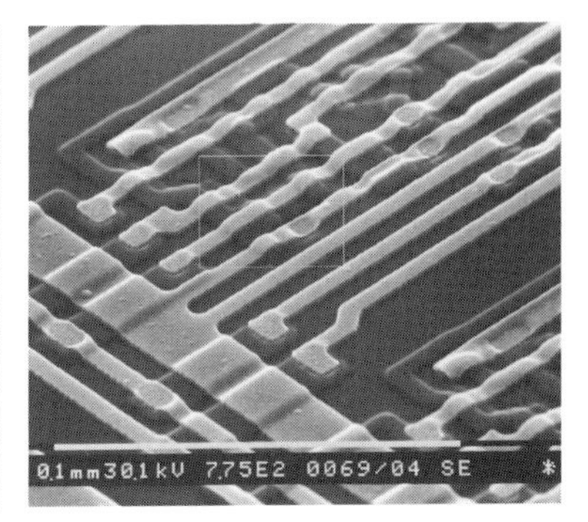

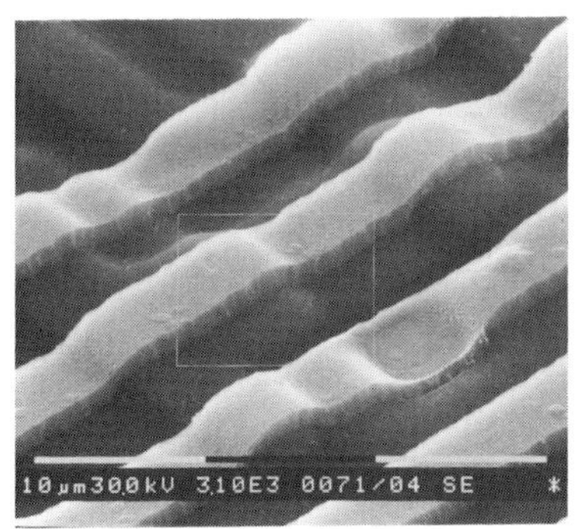

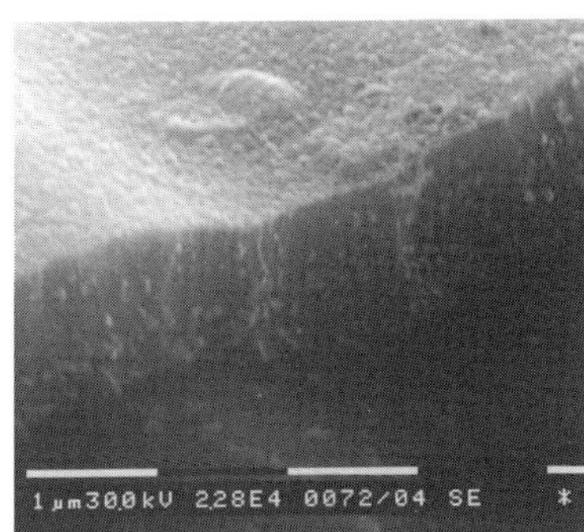

Fig. 6.21 Sequence of electron microscope photographs of a chip with increasing magnification.

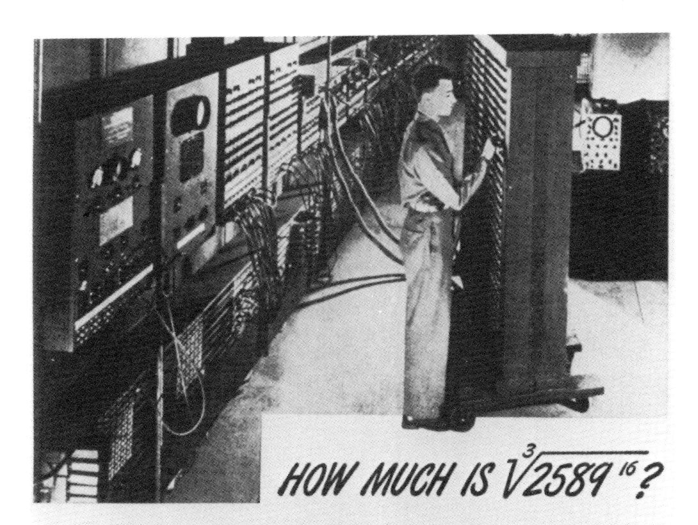

Fig. 6.22 Poster advertising the ENIAC computer. The ENIAC was built to perform ballistic calculations for the US army.

Fig. 6.23 Surrealist poster advertising INMOS chips.

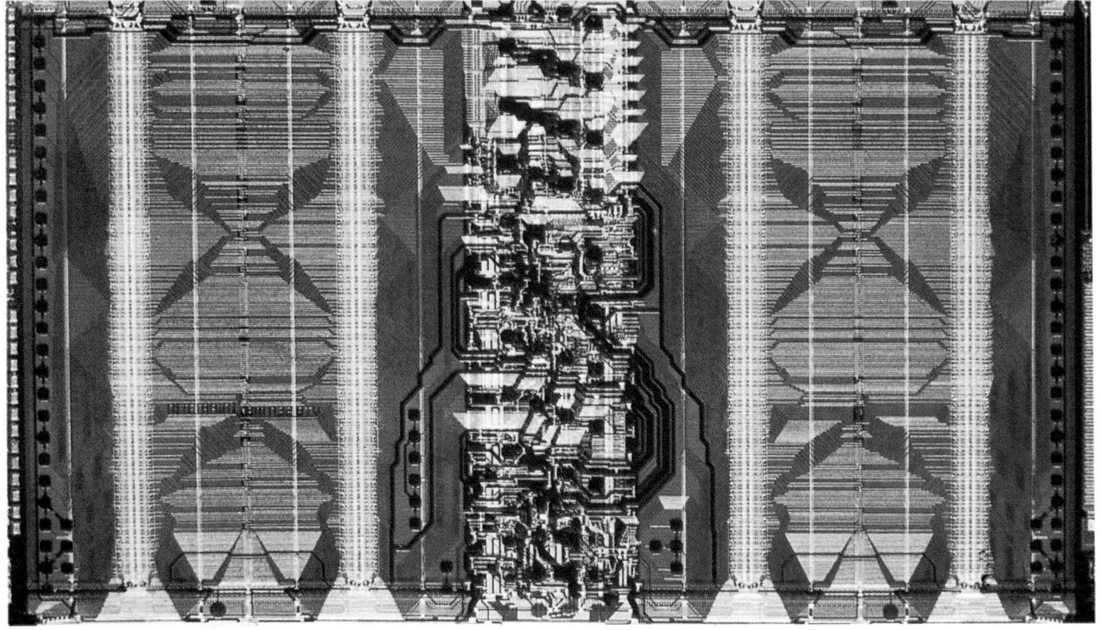

Fig. 6.24 A photograph of a spectacular looking IBM memory chip that can store 294 192 bits of data.

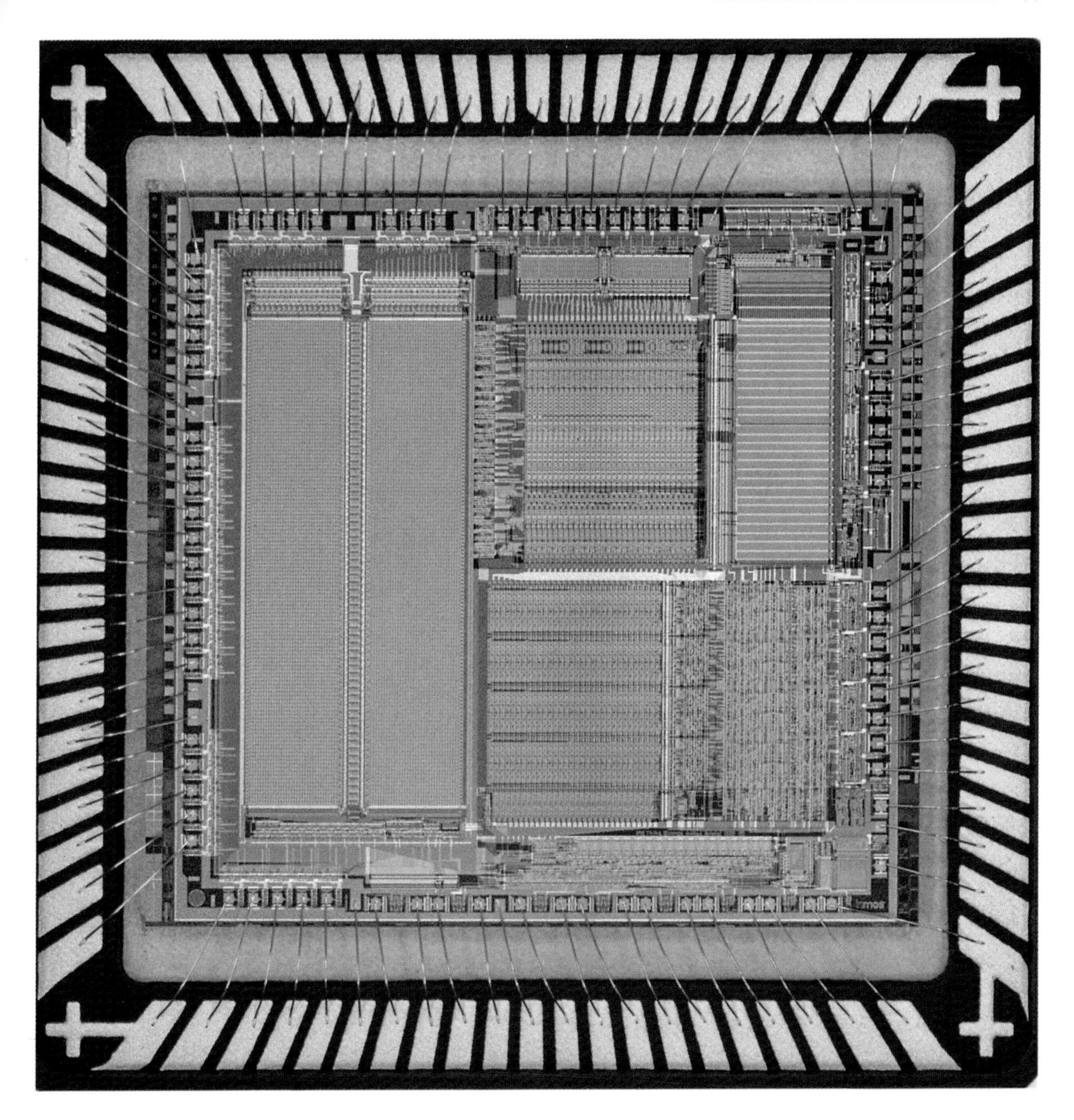

Fig. 6.25 The 'transputer' is a very large scale integration (VLSI) chip containing over 250 000 individual components on about one square centimetre of silicon. It is designed as a building block for computers containing many individual microcomputers but on its own is a very high performance 32 bit microprocessor.

7

Death of a star

One of the most impressive discoveries was the origin of the energy of the stars, that makes them continue to burn. One of the men who discovered this was out with his girl friend the night after he realized that nuclear reactions must be going on in the stars in order to make them shine. She said, 'Look at how pretty the stars shine!' He said, 'Yes, and right now I am the only man in the world who knows why they shine'. She merely laughed at him. She was not impressed with being out with the only man who, at that moment, knew why the stars shine. Well it is sad to be alone, but that is the way of the world.

Richard Feynman

A failed star

In the previous chapter we have seen how quantum mechanics and the exclusion principle provide the basis for an understanding of all the different types of matter we see around us. What is perhaps more surprising is that quantum mechanics and the exclusion principle also provide the key to understanding stellar evolution and the variety of stars. As a prelude to stars we begin with a planet, Jupiter, which in one sense may be regarded as a star that did not quite make it!

Jupiter is by far the largest planet in our solar system. Some impression of its enormous size can be appreciated from the photo-montage shown in fig. 7.2. However, Jupiter is still very much smaller than our star, the Sun. Nevertheless, despite this large difference in size, Jupiter is very similar to the Sun in two important respects. Both are mostly made of hydrogen, and both have an average density only slightly greater than water. Given that they consist of more or less the same ingredients, why then is Jupiter not a fiery burning ball of gas like the Sun?

Imagine descending through Jupiter's cloud tops. As we go down towards the centre, the pressure will increase due to the weight of the overlying gas 'atmosphere' above. Soon the pressure becomes so great that the molecular hydrogen gas is compressed into liquid molecular hydrogen. If we plunge down into this ocean of hydrogen, just like diving in our oceans, the pressure rises still further. The density of the liquid will hardly change, however, since hydrogen molecules have a definite size and the Pauli principle prevents other molecules from encroaching. It is therefore the strength of the covalent hydrogen molecular bonds that is resisting the enormous pressures deep in Jupiter's hydrogen ocean. But as we continue downwards, the pressure rises well beyond anything experienced on Earth. The hydrogen

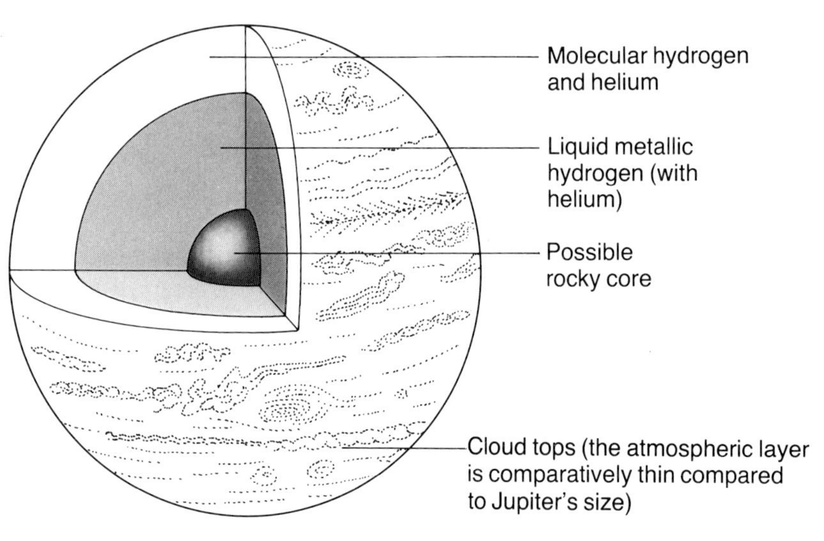

Fig. 7.1 A cut-away model of the interior of Jupiter. At the centre the pressure is 36 million times the ordinary atmospheric pressure on Earth, and the temperature is about 20 000°C. Nonetheless these conditions are not severe enough for Jupiter to be a star. Failed stars like Jupiter are called 'brown dwarfs'.

molecular bonds break and the ocean is now made up of atomic hydrogen. The atoms in this atomic hydrogen liquid are now close enough for a band structure to develop. Since hydrogen has only one electron in the 1S shell, this atomic hydrogen sea will behave like a liquid metal, similar to the liquid metal mercury that we know on Earth. The sea can therefore sustain large electric currents and this is believed to be the key to an understanding of Jupiter's large magnetic field.

The pressure continues to rise as we press on to the centre of Jupiter, but the hydrogen atoms are strong enough to resist the enormous pressures that are generated. What prevents the hydrogen atom from breaking up? It is the familiar electrical attractive force between an electron and a proton that can resist the gravitational pressure of a huge planet like Jupiter. In what sense is Jupiter like a failed star? Stars are very similar to Jupiter except they are a bit more massive. This means that the pressure at the centre of a star will be even greater than at the centre of Jupiter. In a star, this pressure is so great that the electrons and protons in atoms are squeezed apart. Thus, the gravitational forces in a star are strong enough to overcome the electrical attraction between electrons and protons. The resulting soup of electrons and protons is called a 'plasma'.

Planets are held up by atoms. In stars, the atoms have been squashed and the star would like to collapse in upon itself. As the plasma is compressed further, the electrons and protons move about faster and faster and the plasma heats up. This heating and thermal motion of the electrons and protons provide a pressure that tends to oppose further gravitational collapse. However, since the star is able to radiate energy in the form of photons, the plasma will eventually begin to cool. To prevent the star undergoing further collapse, there must be a continuous supply of heat from within the star. As the star collapses, conditions at the centre of the star will eventually become dense enough and hot enough for nuclear reactions to take place. To show that famous physicists are human and make mistakes like anyone else, it is amusing to quote Rutherford on the prospects for nuclear power. He once said 'The energy produced by the breaking down of the atom is a very poor kind of thing. Anyone who expects a source of power from the transformation of these atoms is talking moonshine'. Moonshine or not, nuclear energy is the power that makes the stars shine!

Sir Arthur Eddington (1882–1944) was of Quaker origin and remained a convinced Quaker all his life. His religious beliefs qualified him as a conscientious objector during the First World War. Eddington was one of the first physicists to recognize the importance of general relativity. Indeed, he participated in the 1919 eclipse expedition, which confirmed Einstein's prediction of the bending of light by the Sun. Eddington's major work concerned the theoretical investigation of the interior of stars. He was also a celebrated popularizer of astronomy and was knighted in 1930.

Hydrogen burning

For centuries astronomers and physicists had wondered what made the stars shine. Elementary calculations had shown that ordinary chemical 'burning' was hopelessly inadequate to provide energy throughout the thousands of millions of years of a star's life. The energy had to come from nuclear reactions. It was therefore unfortunate that the famous British astronomer, Sir Arthur Eddington, had shown that the temperatures inside stars were too low for protons to overcome the repulsive energy barrier and get together so that nuclear reactions could take place! Nonetheless, Eddington remained convinced that nuclear energy was the only possible fuel for stars, and he challenged the doubters with the words: 'We do not argue with the critic who urges that stars are not hot enough for this process; we tell him to go and find a hotter place'. Eddington turned out to be right – but quantum mechanics was required to provide the explanation. Using Gamow's quantum tunnelling that we described in chapter 5, a British astronomer, Robert Atkinson, and an Austrian physicist, Fritz Houtermans, solved the problem of energy production in stars. Their paper began with the words: 'Recently Gamow demonstrated that positively charged particles can penetrate the atomic nucleus even if traditional belief holds their energy to be inadequate'. They proposed that light nuclei could

Fig. 7.2 A montage showing the Earth set against the backdrop of Jupiter's cloud tops on the same scale. The giant red spot is visible in the top right of the picture.

act as a 'trap' for protons, and when four protons had been captured an alpha particle could be formed. This could then be ejected from the nucleus thereby liberating a large amount of nuclear binding energy arising from the fusion process in which the four hydrogen nuclei are converted into helium. Their original paper was entitled 'How can one cook helium nuclei in a potential pot?', but this was changed to a more conventional one by the editor of the scientific journal! This paper provided the basis for modern theories of stellar 'thermonuclear' reactions, and ten years later, in 1939, Hans Bethe proposed the so-called 'carbon cycle' in which carbon plays a role similar to Atkinson's and Houtermans' 'proton-trapping' nucleus.

The Sun contains hydrogen and its energy must therefore come from nuclear reactions involving the fusion of hydrogen to make helium and heavier elements. The energy released by a hydrogen bomb also derives from hydrogen fusion reactions. Why, then, does the Sun not explode like a bomb? In fact, the rate of energy generation in the Sun is so slow that a human-sized volume of the Sun burns up its nuclear fuel at a much slower rate than a human converts food into energy! The reason for the great difference in the rate of energy generation in a bomb and a star is that different hydrogen fusion reactions are involved. A star consists mostly of ordinary hydrogen nuclei, each containing a single proton, whereas the fusion reaction used in bombs requires the presence of the two rare isotopes of hydrogen, deuterium and tritium, which contain one and two neutrons, respectively, in addition to the proton. These isotopes of hydrogen undergo nuclear reactions with comparative ease: the nuclear reaction involving

Hans Bethe was born in Strasbourg in 1906 which was then part of Germany. When Hitler came to power, Bethe left Germany and, after a brief stay in England, went to Cornell University in the USA. He was an important participant in the development of nuclear weapons and was also a negotiator for the control of nuclear tests in Geneva, Switzerland. Bethe won the 1967 Nobel Prize for his work on the nuclear processes that make the stars shine.

ordinary hydrogen by which our Sun generates its energy happens so rarely that it has never been observed in the laboratory! This is because the Sun's basic nuclear reaction involves the same mechanism as in a nuclear beta decay. Such reactions are known as 'weak interactions' and proceed very slowly compared with the relatively rapid rates of the 'strong' nuclear interactions, such as deuterium–tritium fusion.

The weak interaction is the name we give to the force responsible for beta radioactivity. The simplest example of a weak interaction is 'neutron beta decay'. Neutrons are more massive than protons and, left to themselves, will eventually 'decay' to a proton and an electron. These two particles are sufficient for conservation of the electric charge – you start with a neutron with no charge and end up with two particles of opposite charge – but experiments indicated that momentum and energy could not be conserved unless there was another, neutral particle also involved. This idea was suggested by Pauli in 1931 – the year before Chadwick discovered the neutron. To distinguish 'Pauli's neutron' from Chadwick's, Fermi called this hypothetical particle a 'neutrino' ('little neutral one' in Italian). Since these curious particles have no charge, they do not feel electric forces. Moreover, since all early attempts to catch a neutrino were unsuccessful, it is evident that they do not feel the nuclear force either! Nevertheless, since they are produced by the weak force, neutrinos must be able to interact with nuclear matter via a weak interaction. The only difficulty is the predicted rate for such neutrino reactions – a neutrino would have to pass through many 'light-years' of matter before it had a fifty–fifty chance of interacting. Since the velocity of light is 300 million metres per second and a light-year is the distance that light would travel in a year (some 30 million seconds) you either have to have a fantastic amount of matter or an enormous number of neutrinos to have a hope of seeing such a neutrino reaction. It is therefore not surprising that it was not until 1956, 25 years after the neutrino was first proposed by Pauli, and long after physicists had accepted that neutrinos must exist, that two US physicists, Frederic Reines and Clyde Cowan, were able to detect weak interactions caused by neutrinos. How did they get enough neutrinos? Since every nuclear fission gives rise to an average of about six beta decay processes, their first idea was to use neutrinos released by a nuclear explosion! Fortunately they were able to use the neutrinos produced by a nuclear reactor instead. Out of the huge number of neutrinos that escape from the reactor – more than a million million neutrinos cross a square centimetre per second – about three neutrino events per hour were observed. The basic neutron beta decay reaction may be written as follows:

$$\mathrm{n} \rightarrow \mathrm{p} + \mathrm{e}^{-} + \bar{\nu}$$

where the particle represented by the Greek letter ν with a bar over the top (and pronounced 'new bar') that is created in the decay is, by convention, taken to be an antineutrino – the antiparticle of the neutrino. If we move one of the participating particles to the other side of such a nuclear reaction equation, to keep the charge balanced and so on we must take it to be an antiparticle. Thus, a possible 'weak interaction' is the process

$$\mathrm{n} + \mathrm{e}^{+} \rightarrow \mathrm{p} + \bar{\nu}$$

which involves the antiparticle of the electron – the 'positron'. In fact, the reaction that Reines and Cowan actually looked for was the reverse of this reaction, namely

$$\bar{\nu} + \mathrm{p} \rightarrow \mathrm{n} + \mathrm{e}^{+}$$

The news of the discovery of the neutrino reached Pauli not long before he died. Nowadays, at the large particle accelerator laboratories, we have lost our sense of wonder about these peculiar particles, and artificially prepared beams of neutrinos and antineutrinos are commonplace. We can therefore now observe neutrino reactions such as

$$\nu + n \rightarrow p + e^-$$

as well as the antineutrino reaction above. We will say more about antiparticles later.

In the Sun there are protons, and protons on their own cannot change into neutrons via the beta decay process

$$p \rightarrow n + e^+ + \nu$$

since neutrons are more massive. However, inside a nucleus this process can, and does, occur if the new nucleus that would be produced by this 'decay' of one of its protons is more tightly bound than the original nucleus. This is because, according to the uncertainty principle, the whole system can 'borrow' the extra energy to make this decay possible, since, at the end of the day, the system will have reached a state with lower energy. Thus, although protons by themselves cannot change into neutrons, if they are in a suitable nucleus they can! This is the key to an understanding of the mechanism by which the Sun produces its energy. Consider the collision of two protons inside the Sun. The electrical repulsion makes it difficult for the two protons to get close enough together to feel the effect of the strong force. But, occasionally, due to quantum tunnelling, two protons come together to form an unstable nucleus containing two protons. Usually, after an extremely short time, the two protons will fly apart again. However, because of the weak interaction and the uncertainty principle, there is a very small chance that one of the protons in this unstable nucleus will beta decay to a neutron and hence result in the formation of a deuterium nucleus:

$$p + p \rightarrow d + e^+ + \nu$$

On average, it takes any given proton inside the Sun more than a thousand million years of collisions before this happens! This very difficult first step is the secret of the slowly burning Sun. Once deuterium has been made, the remaining nuclear reactions required to make helium proceed much more rapidly. These are a strong and electromagnetic interaction between p and d to form ^3He:

$$p + d \rightarrow {}^3He + \gamma$$

followed by a purely strong interaction to produce ^4He:

$${}^3He + {}^3He \rightarrow {}^4He + p + p$$

This sequence of reactions is called the 'proton–proton cycle', and such reactions are believed to be the main source of our Sun's energy. In many stars, however, the temperatures are hot enough for energy generation to proceed via Bethe's carbon cycle. This does not require a weak interaction to take place during the instant of collision. Bethe's mechanism relies instead on the presence of carbon nuclei to act as a sort of catalyst to 'cook' helium.

Despite this stunning triumph of physicists in explaining the origin of the Sun's energy, there remains a niggling little problem that stubbornly refuses to go away, which we should mention as a cautionary postscript. The puzzle is this. The nuclear reactions in the proton–proton cycle of the Sun are believed to be well understood. As we have seen, some of these reactions produce neutrinos and it is relatively straightforward to predict the rate at which such 'solar neutrinos' should arrive at the Earth. An experiment to detect these solar neutrinos has been running for several years in the Homestake Gold Mine in South Dakota. The experiment is set up deep underground to reduce the number of cosmic ray particles from outer space entering the apparatus and causing reactions that could be confused with

Fig. 7.3 The neutrino 'telescope' of Raymond Davis in the Homestake Gold Mine in South Dakota, USA.

Fred Hoyle's ideas on stellar evolution, developed with Geoffrey and Margaret Burbidge and William Fowler in 1957, provide the basis for the theory of supernovas. He was a prominent champion of the 'steady-state' model of the universe which was the main alternative in the 1950s and 1960s to the now generally accepted Big Bang theory. Hoyle has also written several excellent popular books on astronomy as well some very entertaining and perceptive science fiction. Ever controversial, his main research interest at present is the idea that life and diseases come to us from space. He now lives and works in the Lake District in the north of England.

solar neutrino interactions. Unfortunately, even after much careful checking, only about one-third of the expected number of neutrinos are seen. Physicists are still pondering over this problem!

Red giants and white dwarfs

A star like our Sun has enough fuel to keep burning for several thousands of millions of years. But what happens when the hydrogen begins to run out? Since nuclear reactions occur at the core of the star, the core will eventually consist mainly of helium. Helium needs higher temperatures and pressures than hydrogen before nuclear reactions can take place. Thus, the star begins to generate less and less energy and therefore starts to collapse. This collapse causes the temperature to rise until much more rapid hydrogen burning by Bethe's carbon cycle becomes possible. These hydrogen nuclear reactions initially take place in a thin shell about the core. The increased heat production causes the outer layers of the star to expand until its radius is hundreds or thousands of times larger than before. Since the total energy produced by the star must now be spread out over a much larger area, the surface of the giant star is cooler and thus appears redder. The star is now a 'red giant', and in this phase of its development our Sun will be large enough to engulf both Mercury and Venus.

What happens in the core? The core continues to collapse and becomes more and more massive as more helium is produced by the hydrogen burning shell. As the pressure increases, the electrons are crowded closer and closer together. The Pauli exclusion principle does not allow two electrons with the same quantum numbers to occupy the same volume. The size of this minimum volume is determined by the de Broglie wavelength of the electron, and since this wavelength becomes smaller as the momentum of the electron increases, the electrons move faster and faster as the pressure increases. For a star of the mass of our Sun, when the electrons are moving with velocities close to the speed of light, the Pauli principle applied to

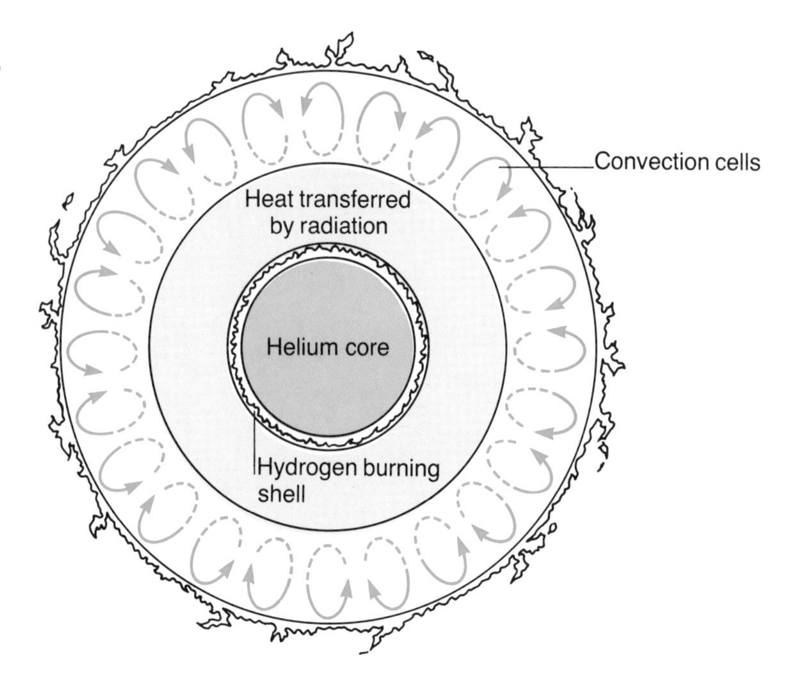

Fig. 7.4 The interior of a red giant star. The diagram is not to scale as the core is small and very dense. The outer layers are extremely tenuous and would occupy a region 100 times the diameter of our Sun.

electrons is able to prevent any further collapse of the core. There will be a similar Pauli principle effect arising from the protons and neutrons in the core. However, because protons and neutrons have a much greater mass than electrons, their de Broglie wavelengths are much smaller and they will not become significant in preventing collapse until much higher pressures are attained. It is therefore the electrons and the Pauli principle that are responsible for preventing further collapse of the core. The core now consists of matter at an incredibly high density – a teaspoonful of this core material would weigh several tons!

Fig. 7.5 This painting shows the Sun as it might look about 5000 million years into the future. The Sun has become a red giant and will eventually engulf the Earth.

Fig. 7.6 At the end of their red giant phase, stars like the Sun will eventually shed their outer layers to form a planetary nebula. The central core that remains will cool to become a white dwarf. This photograph shows the planetary nebula M27 which was ejected from its star some 50 000 years ago. Intense ultra-violet light from the central star keeps the nebula glowing.

For a star like our Sun, the temperature and density in the core will become high enough to allow helium burning to take place. Helium reactions occur rapidly until a hot central core of carbon is formed and the outer layers of the star are thrown off into space. A ring nebula, which is really an expanding shell of gas from such a star, is shown in fig. 7.6. It is kept glowing by radiation from the central core of the star. This central core cools down to become a 'white dwarf': a hot, dense object prevented from further collapse by the electrons and the Pauli principle. A typical white dwarf would be about the size of the Earth yet contain about the mass of our Sun. The white dwarf is white because it is still very hot and therefore able to radiate light energy. However, as no more nuclear reactions are possible, it will gradually cool and grow dim until it becomes a 'black dwarf'. This cooling process is expected to take perhaps a million million years, longer than the present age of the universe, so no black dwarfs have ever been observed.

More complicated evolutions are possible for white dwarf members of a double star system. Fig. 7.7 shows Sirius and its white dwarf companion. Here, it is believed that when the white dwarf was in its red giant phase, material was transferred to its companion star. If the two stars are

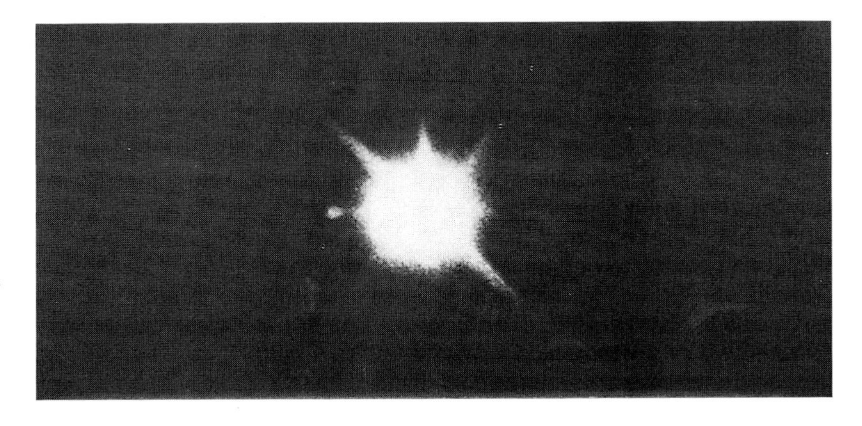

Fig. 7.7 A photograph of Sirius, the brightest star in the night sky, together with its white dwarf companion. The dwarf star is actually hotter than Sirius, but because it is only about the size of the Earth it is much less brilliant.

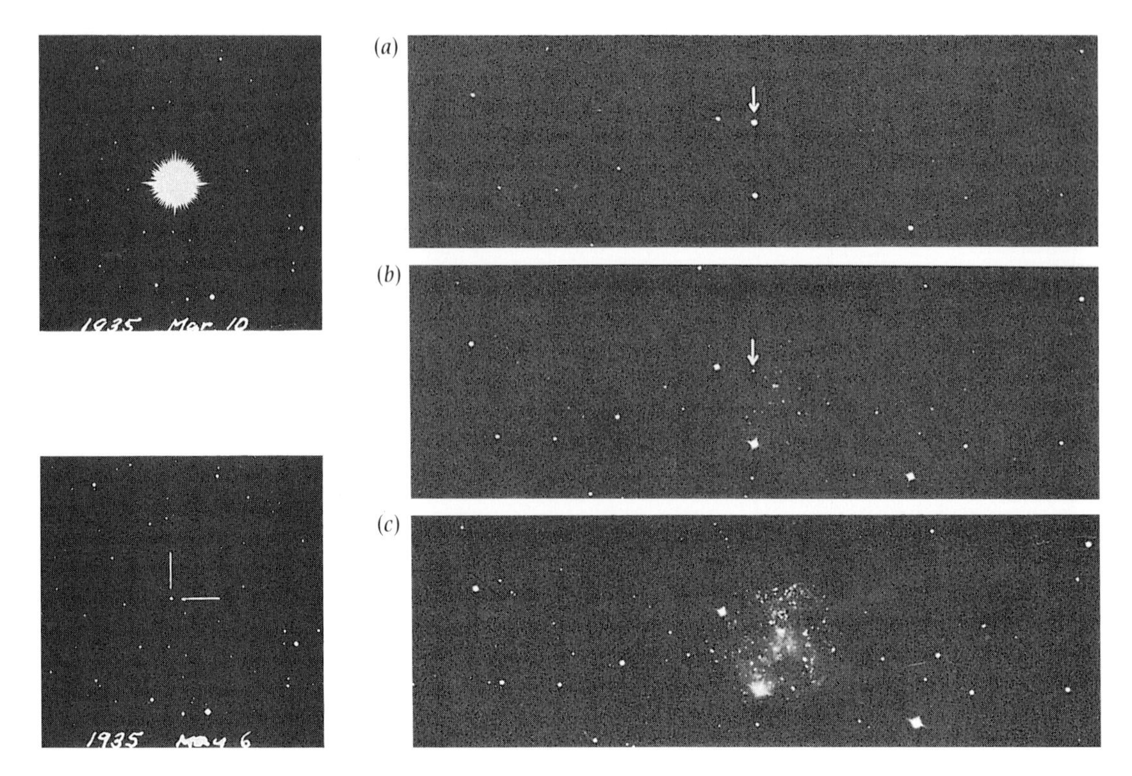

Fig. 7.8 The top photograph shows the nova that exploded in 1934 in the constellation of Hercules as it appeared some months later, in March, 1935. The bottom photograph shows the nova as it appeared eight weeks later.

Fig. 7.9 Three photographs of a supernova in the galaxy IC4182. (a) In this short 20 minute exposure, taken on August 23rd, 1937, the supernova shows up clearly while the galaxy is too faint to be seen at this exposure. (b) On this photograph, taken just over a year later on November 24th, 1938, the supernova is much fainter. With this 45 minute exposure it can just be seen and the galaxy is starting to become visible. (c) On January 19th, 1942, the supernova is too faint to be seen. This 85 minute exposure shows the galaxy clearly, and this series of photographs emphasizes the extreme brilliance of supernova explosions. Over 400 supernovas have been recorded but the last one in our Galaxy was observed by Kepler in 1604.

sufficiently close, matter can also be transferred to a white dwarf from a close red giant. The white dwarf can accumulate hydrogen in this way and this can result in a violent nuclear explosion. For a brief period of time, the double star system can become several tens of thousands of times brighter. In the days before telescopes it seemed as if a new star had been born, which then faded and died in a matter of weeks. Such stars are called 'novas', from the Latin word for new.

A white dwarf in a double star system may also be involved in the most violent of all star explosions – a 'supernova'. In this case, the new star may appear as bright as a whole galaxy. No supernovas have been observed in our own Galaxy since the telescope has been invented, but, in 1054, the Chinese recorded the appearance of a 'guest star' that was bright enough to be visible in the daytime for many days. At the position of this explosion, we now see the spectacular Crab nebula, which certainly looks like the remains of some enormous explosion. In fact, if the star is sufficiently massive, supernovas can occur without the aid of a companion star. The Crab supernova observed by the Chinese is believed to have been of this type. In the next section, we shall see that quantum mechanics and the Pauli

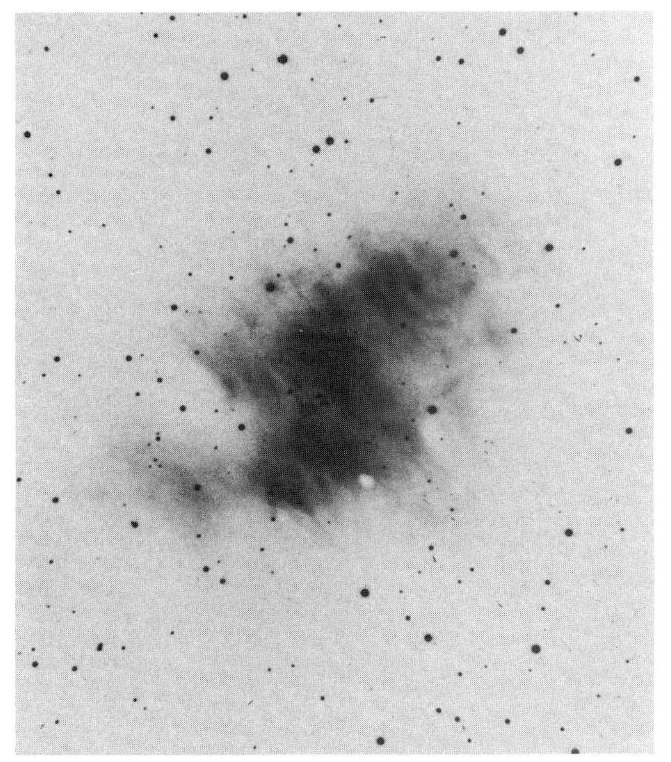

凡十一日沒三年三月乙巳出東南方大中祥符四
年正月丁丑見南斗魁前天禧五年四月丙辰出軒轅
前星西北大如桃速行經軒轅大星入太微垣掩右執
法犯次將歷屏星西北凡七十五日入濁沒明道元
年六月乙巳出東北方近濁有芒彗至丁巳凡十三
日沒至和元年五月己丑出天關東南可數寸歲餘
稍沒熙寧二年六月丙辰出箕度中至七月丁卯犯
箕乃散三年十一月丁未出天囷元祐六年十一月
辛亥出參度中犯掩側星壬子犯九游星十二月癸
酉入奎至七年三月辛亥乃散紹興八年五月守婁

Fig. 7.10 The Crab nebula is formed from the remains of a star that was seen to explode by the Chinese. Two hours after midnight in the summer of 1054, Chinese astronomers noted the appearance of a 'guest star' between the horns of the constellation we call Taurus – the Bull. It outshone Venus and Jupiter and was even visible during the day for about three weeks. The Chinese historical record of this event is shown together with a photograph of the Crab nebula as it appears today. At the centre of the Crab nebula there is a 'pulsar' – a rapidly rotating neutron star formed in the explosion.

principle have still more to say about the evolution of such very massive stars into even more exotic objects than white dwarfs. The range of applicability of quantum mechanics is truly breathtaking!

As a footnote to this section, we ought to complete our family of dwarfs with a mention of 'brown dwarfs'. These are believed to be failed stars, rather like Jupiter, in which nuclear reactions never really got going. Brown dwarfs are currently exciting much interest amongst astronomers.

Neutron stars and black holes

In very massive stars, helium burning to carbon is not the final chain of nuclear reactions. The core is sufficiently hot that new nuclear processes can occur. A very complicated series of reactions can make successively heavier elements up to iron, ^{56}Fe. After iron, it is not possible to gain more energy from fusion reactions producing yet heavier elements, since iron has the highest binding energy of all the elements. More and more iron will therefore accumulate in the core until the nuclear fuel begins to run out. As before, the core then contracts until it is prevented from further collapse by the exclusion principle for the electrons in the white dwarf material of the core.

Can the electron Pauli principle prevent the collapse of any star, no matter how massive? The answer is no. There is a critical mass – called the Chandrasekar limit – beyond which the Pauli principle applied to electrons cannot prevent further gravitational collapse to an even more exotic and denser form of matter than in a white dwarf star. How does this come about? As the iron core of a very massive star collapses, the electrons eventually get squeezed together so much that a significant number of them have enough energy to initiate a weak interaction process. This changes protons into neutrons via the reaction

$$e^- + p \rightarrow n + \nu$$

Fig. 7.11 More than 2000 dipole antennae made up the radio telescope with which pulsars were discovered. The sloping bars support a reflecting screen that increases the telescope's sensitivity. The original purpose of the instrument was to study the twinkling of radio sources.

This has the effect of removing electrons and protons from the core, as well as allowing energy, in the form of neutrinos, to escape from the star. Once this process starts to reduce the Pauli 'pressure' due to the electrons, an incredibly rapid and violent collapse of the core is initiated. The precise details of this collapse and how the spectacular supernova explosion is generated are very complicated and still debated by astrophysicists. What now seems clear, however, is that the supernova will leave behind a compressed ball of hot neutrons – a 'neutron star'. As the hot neutron star cools, any further collapse is prevented by the Pauli principle applied to the neutrons, unless the mass is so great that the star can become a 'black hole', as we discuss later. For a star remnant about twice as massive as our Sun, the resulting neutron star will be about ten miles in diameter. The density is over a million million times that of water and is roughly the same as that inside an atomic nucleus. In a sense, therefore, a neutron star is just like a gigantic nucleus!

Subrahmanyan Chandrasekar was born in Lahore, India (now Pakistan) in 1910, and was educated at Madras University. He obtained a PhD at Cambridge, UK, where he studied under Dirac. He then worked at Chicago University and the Yerkes Observatory. Chrandrasekar developed the first consistent model of white dwarf stars and was awarded the Nobel Prize in 1983 with William Fowler.

Neutron stars may sound like a very fanciful application of quantum mechanics, but they were in fact suggested nearly 50 years ago by J. Robert Oppenheimer. Oppenheimer is such an interesting character, and one who played such a pivotal role in the development of our present MAD world (where MAD stands for mutually assured destruction) that we cannot resist a brief historical digression. The austere, academic scientist who predicted neutron stars was the same man who later directed the physicists in the Manhattan project to make the atom bomb. This was the same Oppenheimer who, in 1954, was declared a security risk and 'unfit to serve his country'. Edward Teller, a colleague of Oppenheimer at Los Alamos during the war and later popularly known as the 'father of the hydrogen bomb', split the scientific community by giving evidence against Oppenheimer. The background to this indictment was turbulent and confused with Senator Joseph McCarthy's witch-hunts against communists reaching their peak. Moreover, Klaus Fuchs, who was co-author with von Neumann of the top-secret document called a 'Disclosure of invention', which contained a summary of every significant advance made towards the fusion thermonuclear bomb, had been caught by British security in 1950 giving bomb secrets to the Russians. To cap it all, in August of 1953 the Russians had succeeded in testing the world's first true, droppable H-bomb, and it would not be until

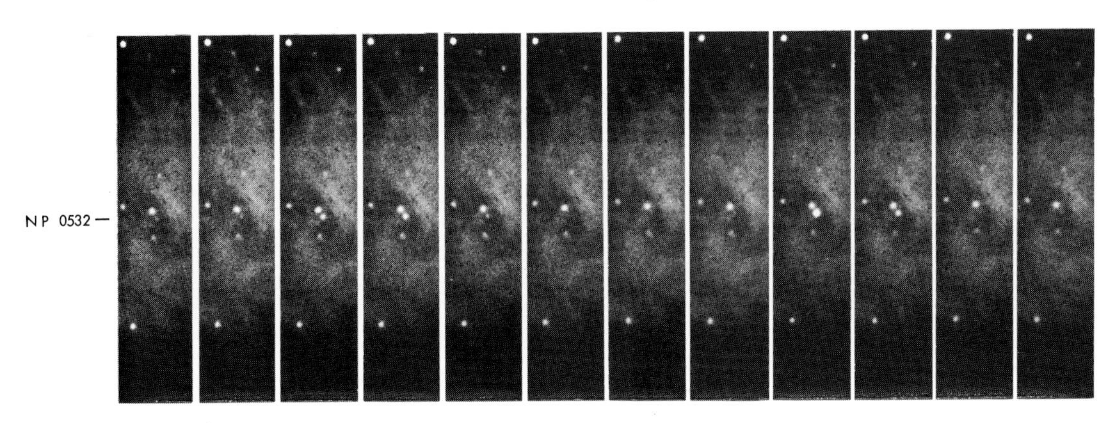

Fig. 7.12 This sequence of photographs shows the Crab pulsar NP0532 flashing on and off for a complete cycle. The whole cycle takes only 1/30th of a second, which is the period of rotation of the neutron star.

1956 that the USA would have a usable fusion bomb. It is perhaps not surprising, therefore, that it was not until the morning of November 22nd, 1963, that the White House announced that President Kennedy would personally present Oppenheimer with the prestigious Fermi award. This was intended as a first step towards a public apology for the anticommunist hysteria that had led to his indictment ten years earlier. On the afternoon of the same day, John Kennedy was assassinated, and it was left for President Johnson, in opposition to warnings from his political advisors, to make a personal presentation of the award. One US senator who had been prominent in the campaign against Oppenheimer called the award ceremony 'shocking and revolting'.

After this historical interlude, we must return to neutron stars and explain why astronomers believe they exist. The observational evidence for neutron stars is associated with the discovery of 'pulsars' by Joycelyn Bell, a research student of Anthony Hewish, in Cambridge, UK, in 1967. Pulsars are rapid and remarkably regular radio 'pulses' of extra-terrestrial origin. Soon after the first pulsar was discovered, one was found at the centre of the Crab nebula, at the site of the Chinese supernova explosion. The Crab pulsar flashes on and off about 30 times a second, emitting energy across most of the electromagnetic spectrum (see appendix 1). Pulsars were originally designated by the acronym LGM – standing for little green men – since they were first suspected to be signals from an extra-terrestrial civilization. The truth is now believed to be somewhat less romantic – they are now thought to be rapidly rotating neutron stars!

Joycelyn Bell Burnell was a graduate student working under Anthony Hewish in Cambridge, UK, when she first noticed the regular signal of pulsars. Her thesis contains results on the angular diameter of some 200 scintillating radio sources and only mentions pulsars in an appendix! Anthony Hewish who developed the scintillation technique and supervised the project won the Nobel Prize in 1974.

It was Tommy Gold of Cornell, USA, who first realized that pulsars could be understood as rotating neutron stars. The rate of rotation required, however, was very much greater than that of normal stars. But, just as a skater draws in her arms to make a slow spin become a rapid spin (an elegant demonstration of angular momentum conservation) so too does the rotation of a star increase as it collapses to form a neutron star. Moreover, the magnetic fields of the star are also increased to a much higher value by this collapse. Now, as shown in fig. 7.13, the magnetic poles will not usually coincide with the poles of the axis of rotation. By a rather complicated mechanism involving both the magnetic and electric fields of the neutron star, it is believed that an intense narrow beam of radiation is produced in the direction of the magnetic axis. It is this beam of radiation sweeping regularly across the Earth as the neutron star rotates that causes the observed 'pulsing' of the pulsar.

A neutron star is an amazingly compact and dense object. Nevertheless, the immense gravitational forces generated within such an object are countered by the neutron Pauli principle. However, it is believed that if the star is sufficiently massive (more than about three times the mass of our sun)

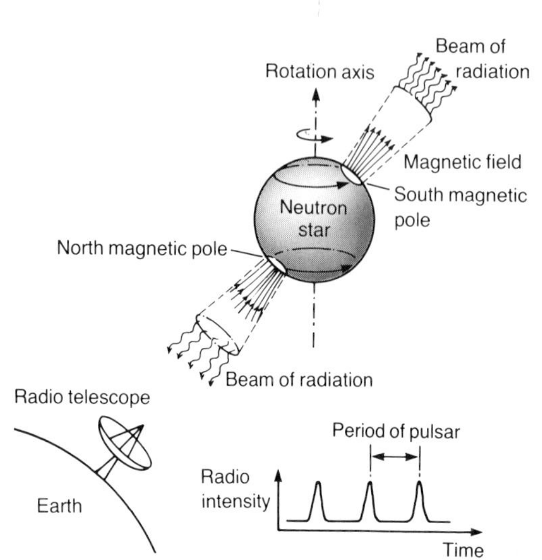

Fig. 7.13 A pulsar is a rapidly rotating neutron star with an enormous magnetic field. As the star rotates, it emits a narrow beam of radiation from the polar regions. If the beam crosses the Earth, the pulsar can be detected by the regular series of pulses of radio energy received.

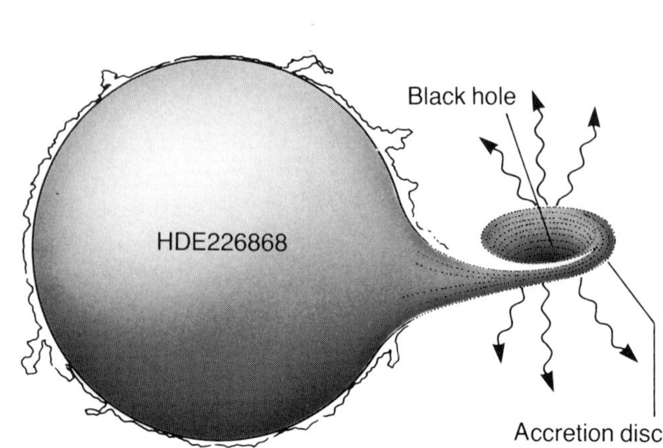

Fig. 7.14 A sketch of a black hole model for the X-ray source Cygnus X1. Measurements of the period of rotation of this binary system suggest that the mass of the unseen X-ray source is larger than the mass of a neutron star. It is suggested that the X-rays are produced as material streaming from the companion star falls onto the 'accretion disc' of material rotating around the black hole, before eventually falling beyond the region of no return.

even the Pauli principle for the quarks inside neutrons (see chapter 10) cannot prevent the star collapsing to form an even more bizarre object, a 'black hole'. Such objects are permitted by Einstein's 'general theory of relativity' and correspond to a special type of solution to Einstein's equations. To form a black hole requires enormously high densities. For example, for our Sun to become a black hole, it would need to be compressed to the size of a ball about four miles in diameter. Once a star has been compressed smaller than a critical radius, the so-called 'Schwarzschild radius', the effects of gravity are so strong that nothing, not even light, can escape. It is truly a black hole!

At present, we have no complete theory that combines both quantum mechanics and general relativity in a totally satisfactory way. Thus, we cannot be sure of the details of a star's collapse to a black hole, or even be absolutely sure that such objects must exist. Our doubts would be removed if we could observe a black hole experimentally. But since no radiation can ever leave the black hole, how could this be done? One way that has been suggested is to look for 'binary star' systems. These normally consist of two stars rotating round each other, like a couple on a dance floor. However, if one of the objects in such a system is a black hole, its mass can be estimated from the behaviour of the visible partner. Moreover, the black hole 'sucks in' matter from the partner star and this material will radiate X-rays – high energy photons – as it falls towards the black hole. A candidate black hole has been seen in the Cygnus constellation, but the question has not yet been settled beyond doubt. We shall have more to say about quantum mechanics and black holes in chapter 9.

8

Quantum co-operation and superfluids

. . . there are certain situations in which the peculiarities of quantum mechanics can come out in a special way on a large scale.

Richard Feynman

Laser light

Nowadays most people have heard of 'lasers', and laser light displays are a frequent ingredient of modern rock concerts. Laser light has many applications ranging from astronomy to hydrogen fusion. What is the special feature of laser light that makes it so useful? The answer to this question involves a property of wave motion known as 'coherence', together with light photons acting together in a special form of quantum mechanical co-operation. This type of quantum co-operation will turn out to be vital for an understanding of the peculiar behaviour of quantum 'superfluids'. To understand the special nature of laser light, however, we must first explain what is meant by coherence.

Consider the simple wave motion shown in fig. 8.2. As we have seen, the pattern repeats after one 'wavelength' and the 'frequency' of the wave corresponds to the number of wavelengths sent out per second. If this wave is a wave on a string, each point on the string just moves up and down with a certain 'amplitude': the maximum distance the point can move out to before it starts to come back. Up to now, this is really all we have needed to know about waves. Now consider two waves of the same wavelength but started at slightly different times, as shown in fig. 8.3. In the first case, both waves have their crests and troughs at the same point. In the second case, the dashed wave has started to fall before the other has reached its peak. The next figure shows the extreme case when one wave has a trough when the other has a crest. This is similar to our discussion of interference in chapter 1. We say that these two wave motions have different 'phase differences' in these three cases. The phase of a wave is what tells us where a point on the wave has got to on its up and down motion. If there is a definite phase difference between two waves, as shown in the figure, the two waves are said to be 'coherent' and they will display the usual interference effects. Two different atomic light sources, on the other hand, do not show interference effects and are said to be 'incoherent'. The reason for the lack of any interference is that the light in the two sources is produced by many different atoms emitting photons at different times. Each lamp therefore sends out light consisting of lots of waves with many different phases. There is therefore no definite phase difference between the light waves coming from the two sources, and all the delicate interference effects are washed out. By contrast, laser light is remarkable in that light from many different atoms is radiated in phase. It is this coherence property of laser light that makes it possible to focus a laser beam on a very small spot and obtain a very high concentration of light energy. A laser beam with less power than an ordinary light bulb can burn a hole in a metal plate.

Charles Townes was born in South Carolina in 1915. During the Second World War he worked for Bell Telephone Laboratories on the design of radar systems. Apparently one morning in 1951, while waiting for a restaurant to open for breakfast, he had the idea of using molecules rather than electronic circuits to generate microwaves – short wavelength radio waves. By 1953 he had built the 'maser' (microwave amplification by stimulated emission of radiation) using ammonia molecules. Townes went on to speculate about the construction of a similar device for visible radiation.

Laser is an acronym standing for 'light amplification by stimulated emission of radiation'. Stimulated emission is a process of interaction of an atom with light that we have not come across so far. We have seen that if we shine light on an atom with a photon energy that corresponds exactly to an energy level difference, the electron is 'stimulated' to jump to the higher

Fig. 8.1 The top photograph shows the Moon as seen from the Lure Observatory on the island of Maui, in Hawaii. A pulse of laser light is fired through the telescope at the Moon. By the time the pulse reaches the Moon it has spread out to cover an area more than two miles wide. Part of the beam is then reflected back to Earth by special reflectors left on the surface of the moon by the Apollo 14 astronauts. The return signal can be picked up and the time the light beam has taken to travel from the Earth to the Moon and back can be accurately timed. The travel time is about 2.5 s, and measurements such as this allow us to determine the distance to the Moon to within an accuracy of a few centimetres. The special lunar reflectors are shown in the bottom photograph.

state. This process is therefore sometimes called 'stimulated absorption'. We also know that an atom in an 'excited' state will 'spontaneously' emit a photon of the right energy and the electron will jump to the ground state. This process of 'decay' of an excited atom is called 'spontaneous emission'. A third type of process involving photons was discovered by Einstein as early as 1916. In November of that year, he wrote to a life-long friend, Michele Angelo Besso: 'A splendid light has dawned on me about the absorption and emission of radiation'. (When Besso died in 1955 Einstein wrote to the family: 'What I most admired in him as a human being is the fact that he managed to live for many years not only in peace but also in lasting harmony with a woman – an undertaking in which I twice failed rather

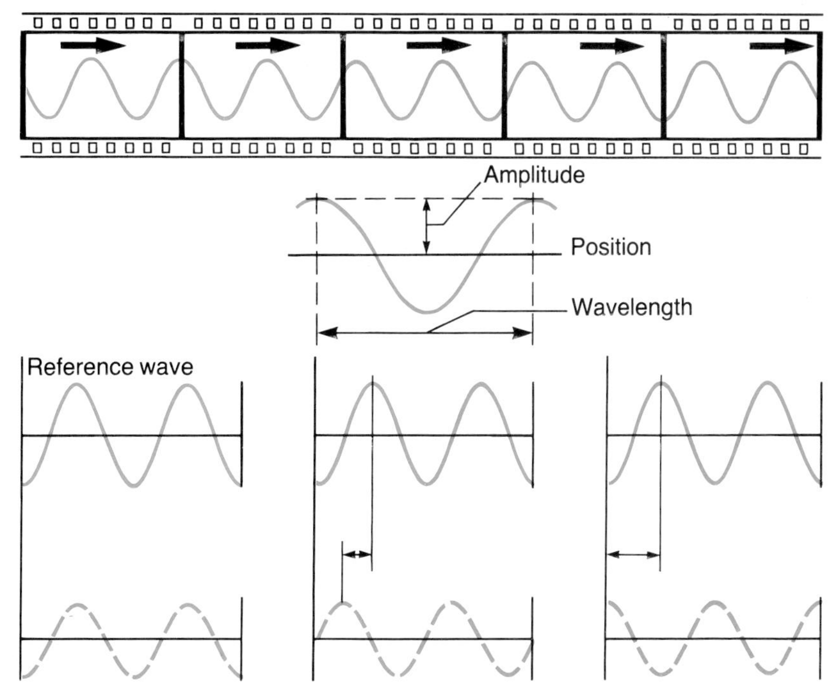

Fig. 8.2 A sequence of 'photographs' of a passing wave. The photographer is standing still – the arrow points over the same crest in all the frames showing how the wave moves to the right. The sketch below shows how the amplitude and wavelength of such a wave motion are defined.

Fig. 8.3 Two waves with the same wavelength but three different start times.

Fig. 8.4 Welding by laser at the Mirafiori car plant of Fiat in Turin. The laser beam, which is invisible, is delivered from the cone shaped nozzle at the end of the welding head just above the sparks. A 2.5 kW carbon dioxide laser generates the beam.

Theodore Harold Maiman was born the son of an electrical engineer in 1927. He paid his way through college by repairing electrical appliances. While working for Howard Hughes' research laboratories he became interested in the maser invented by Townes and the problem of constructing a similar device for light. Maiman constructed the first laser in 1960.

disgracefully'.) Einstein had realized that if light with the correct photon energy for an electron transition is shone on an atom in an excited state, then the atom can be 'stimulated' to make a transition to the lower energy state. It is natural to call this process 'stimulated emission' of radiation. The excited atom would, of course, have made a transition to the lower state sooner or later – it just makes it sooner in the presence of the stimulus of the radiation. For over 35 years this stimulated emission process gained hardly more than a cursory comment in quantum mechanics textbooks, since it seemed to have no practical application. What had been overlooked, however, was the special nature of the light that is emitted in this way. The photons that are emitted have exactly the same phase as the 'inducing' photons. This is because the varying electric fields of the applied light wave cause the charge distribution of the excited atom to oscillate in phase with this radiation. The emitted photons are all in phase – they are 'coherent' – and, furthermore, they travel in the same direction as the inducing photon.

Fig. 8.5 Sequence of figures showing the three possible 'transition processes' for light photons and electrons in an atom. (a) *A diagram illustrating the processes of 'stimulated absorption' and 'spontaneous emission'. In the first case a photon with the right energy can be absorbed by the electron in the ground state and cause it to jump to an excited energy level. After a time this electron will drop back into the lowest energy level giving off a photon with the same energy as the one that was absorbed. This 'decay' process is called spontaneous emission.* (b) *'Stimulated emission' of radiation occurs when photons are directed at an atom which is already in an excited state. The stimulating photon and the photon radiated in the transition have the same energy and the same phase.*

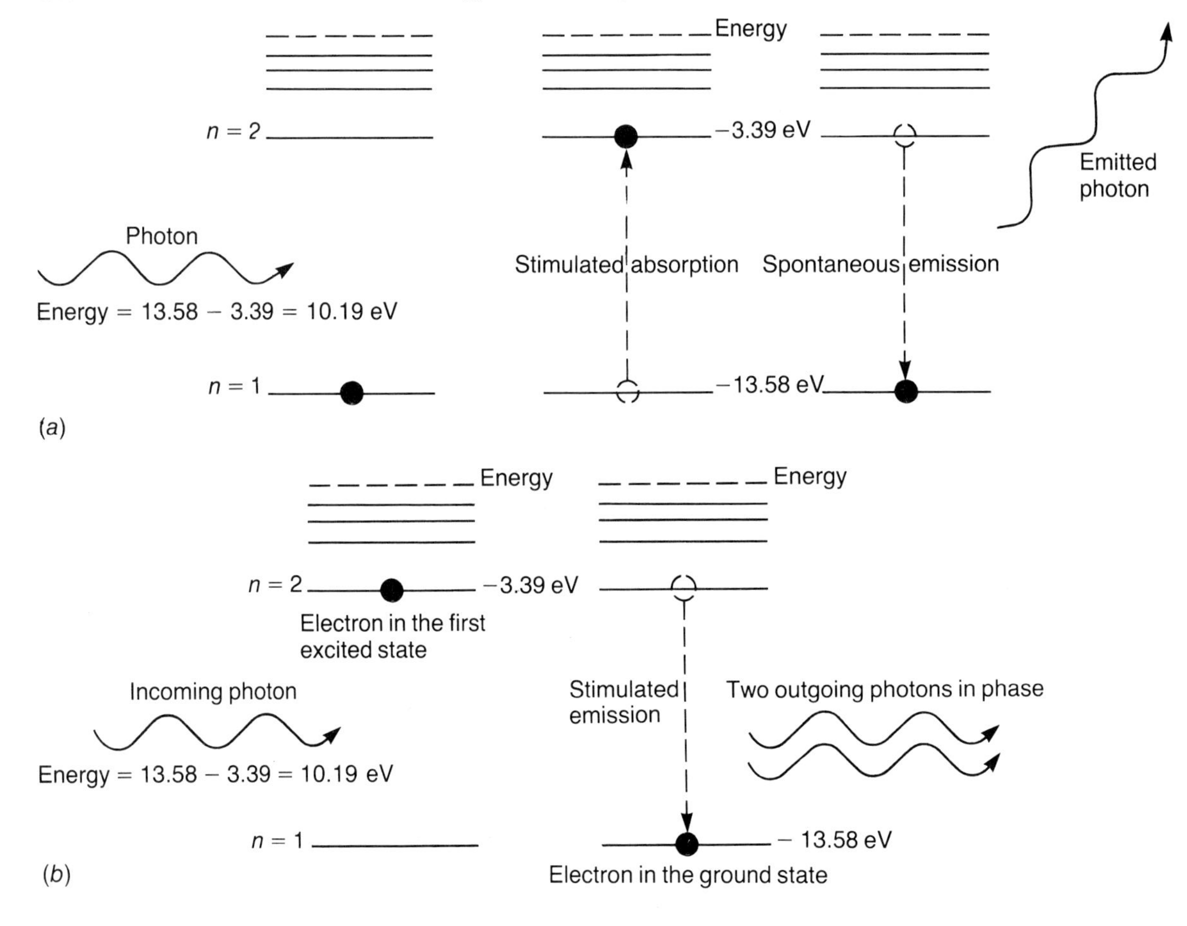

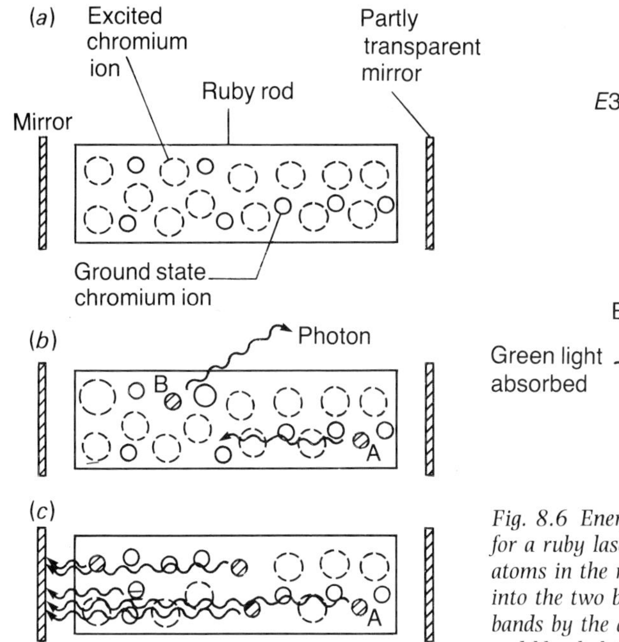

Fig. 8.7 *A sequence of figures showing the build up of laser light.* (a) *This illustrates the state of affairs after a population inversion has been achieved. The smaller circles represent chromium atoms in their ground state and the larger dashed circles the excited atoms.* (b) *In this figure two atoms have decayed back to the ground state. In one case the photon leaves the side of the ruby and cannot cause any stimulated emission. In the case where the photon is emitted along the length of the rod, however, this can generate more and more decay photons, all with the same phase.* (c) *This figure shows how the mirrors at the ends help build up a beam parallel to the length of the rod. Several photons are just about to be reflected by the end mirror and will cause more transitions as they pass back along the crystal.*

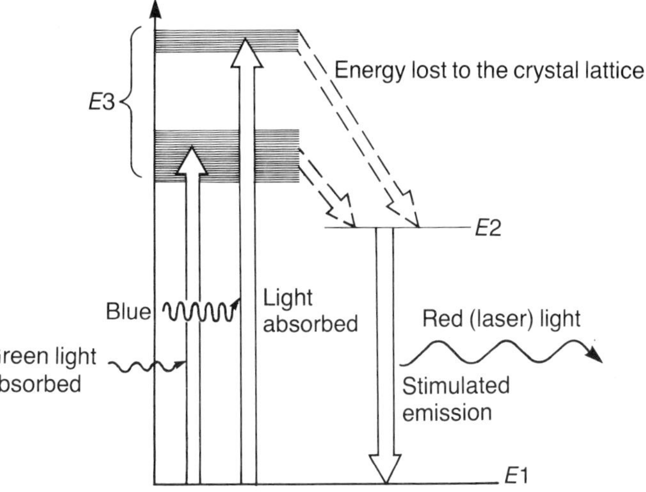

Fig. 8.6 *Energy level diagram for a ruby laser. The chromium atoms in the ruby are 'pumped' into the two broad excited energy bands by the absorption of green and blue light. These excited atoms quickly lose energy to the crystal lattice and the electrons drop into the long-lived 'metastable' energy level shown as* E_2 *in the figure. More electrons will be in this level than in the ground state so a 'population inversion' exists. Stimulated emission of the transition from this level to the ground state results in red laser light.*

This is all very well, but there are a number of technical problems to be solved before we can produce intense beams of laser light. At normal temperatures most atoms are in their ground state. We have to look for a way of 'pumping' energy into the laser material so that we manage to get most of the atoms into an excited state. Having more atoms in an excited state than in the ground state is not a normal state of affairs – it is called a 'population inversion'. If we can arrange for such a population inversion, then the stimulated emission process will exceed the stimulated absorption and we will obtain a net amplification of the stimulating light.

The first laser used a ruby crystal, consisting of aluminium oxide in which some of the aluminium atoms had been replaced by 'impurity' chromium atoms. The relevant energy levels of the chromium atoms in this system are shown in fig. 8.6. By pumping in light with an energy corresponding to the difference between E_1 and E_3 the chromium atoms are excited to the broad short-lived upper state. These excited atoms then decay very quickly to a relatively long-lived state E_2 and a population inversion is obtained. When some of these E_2 states decay spontaneously, the right conditions exist for these photons to cause stimulated emission of the other excited atoms. A diagram of a ruby laser is shown in fig. 8.7. Photons will be emitted in all directions, but those that do not travel along the length of the ruby rod will escape through the sides before they cause much stimulated emission. Photons travelling along the axis of the rod will be reflected back and forth several times by the mirrors at the ends of the rod. Thus, more and more atoms will be stimulated to emit photons and an intense coherent beam of 'laser' light will emerge from the partially reflecting end of the crystal. In this first laser, the pumping required to obtain the crucial population inversion was provided by a flash of light, and a special long-lived 'metastable' state (E_2 in fig. 8.6) was required to maintain the inversion. Modern lasers can be continuously pumped and do not need the lasing state to be specially long-lived.

Fig. 8.8 Interior of a ruby laser. The pink cylinder in the top section is the ruby, while the cylinder in the bottom section is the flash tube used to generate the population inversion. Both are cooled by water which enters the apparatus through the pipes visible in the photograph. The mirror on the left is fully reflecting while the one on the right is partly transparent to let the laser beam emerge.

The fact that many photons in a laser beam are in the same quantum state is possible only because photons are 'bosons'. For electrons, Pauli does not allow more than one to have the same quantum number, but photons actually prefer to be all together in the same quantum state. We shall discuss this property of bosons in more detail in the next section. We end this section on lasers with a mention of two very different applications of laser light.

The unique properties of laser light enable us to concentrate light energy in a very intense, short pulse. Using such laser beams, the distance of the Moon from the Earth can be measured with amazing accuracy. Fig. 8.1 shows the footprints of the Apollo 14 astronauts near a special reflector they had placed on the Moon's surface. A pulse of laser light is fired through a large telescope at the Moon. By measuring the time of flight of photons reflected from the Moon, the distance to the Moon can be determined to an accuracy of within a few centimetres in a distance of about 400 000 km.

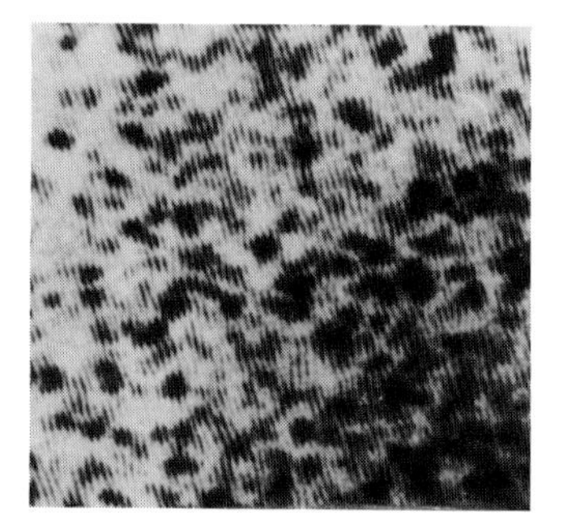

Fig. 8.9 The smudgy looking photograph at the top left is a hologram. The other three photographs show three different views all generated by the same hologram. Not only can one see round objects in different views but also the same pictures, albeit somewhat degraded in clarity, can be produced from just a small piece of the original hologram. This is possible because of the interference mechanism that underlies holography.

Dennis Gabor (1900–79) was born in Budapest but received his education in Germany. When Hitler came to power, he came to England and worked as a research engineer for the Thompson–Houston electrical manufacturing company in Rugby. His original paper on holography in 1948 was in the context of electron optics, and it was not until the invention of lasers that his idea gained much attention. He was awarded the Nobel Prize in 1971.

Another interesting application of laser light is to three-dimensional photography or 'holography'. Light from a laser is split into two beams by a half-silvered mirror. One beam illuminates the object and the scattered light falls on a photographic plate. The other beam is directed at the photographic plate without scattering off the object. Since the laser light is coherent, the two beams can interfere. Thus, the photographic plate records the interference pattern caused by the recombination of these two beams. This photographic record of the interference pattern is called a 'hologram', after the Greek word *holos*, meaning 'whole'. This is because a hologram, unlike an ordinary photograph, which just records the intensity of the light falling on the photographic plate, also contains information about the phase of the scattered light, since it is a record of an interference pattern. It therefore contains the whole of the optical information coming from the object being photographed. A hologram bears no resemblance to the object being photographed – it looks like an almost random pattern of smudgy dots. However, when the hologram is illuminated by a beam of laser light, a perfect three-dimensional copy of the original object is reconstructed. If you walk around and look at the image from different angles you see the relative positions of the various pieces exactly as you would for the real thing. In particular, things hidden from view in one position can be seen by looking at the holographic image from another direction. Holography was invented by Hungarian-born Dennis Gabor working in Rugby, UK, in 1947, but remained merely a 'scientific curiosity' for about 15 years. It is only since the advent of coherent beams of laser light that holography has become a multi-million dollar industry, with applications ranging from medical diagnosis to tyre testing.

Bose condensation and superfluid helium

A photograph of Satyendra Bose (1894–1974). After his paper on the quantum theory of light had been rejected for publication, Bose sent a copy to Einstein in desparation. Einstein personally translated Bose's paper from English into German and arranged for it to be published. From being virtually unknown, Bose suddenly became an internationally famous physicist.

We have seen, in chapter 6, how Pauli's exclusion principle, applied to atomic electrons, is able to explain Mendeleev's periodic table of the elements. All the basic 'matter-like' particles – electrons, protons and neutrons – obey the Pauli principle. No two identical 'fermions', as these particles are called, can occupy the same quantum state. Thus, if we consider putting electrons in a box potential, the electrons cannot all go into the lowest energy level. Instead, they fill up the quantized energy levels in pairs with opposite spin, so that no two electrons have the same quantum numbers. This is what happens for 'matter-like' particles. However, 'radiation-like' particles, such as photons, behave very differently. These particles actually prefer to be in the same state! Such particles are called 'bosons' in honour of an Indian physicist Satyendra Bose.

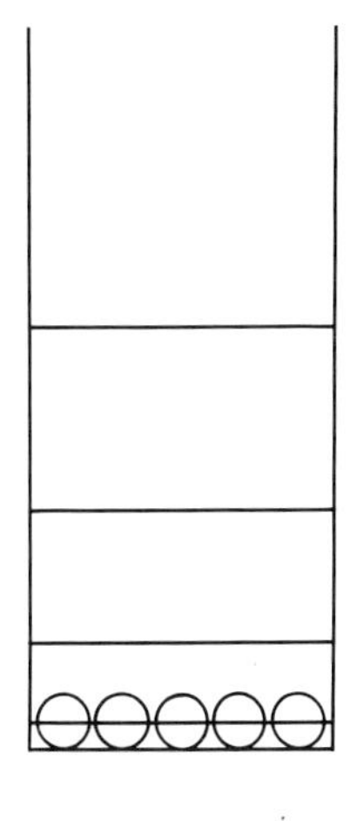

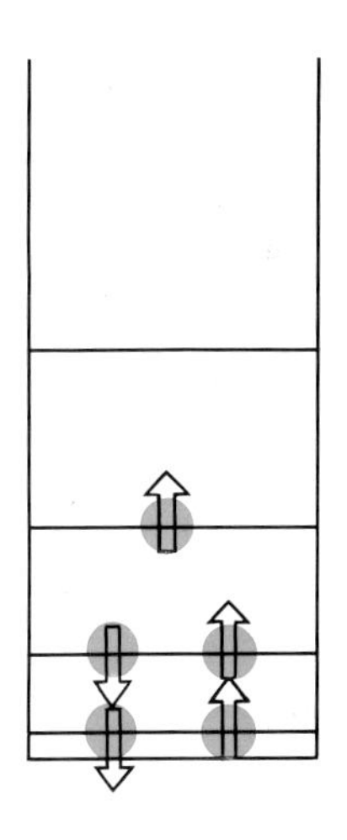

Fig. 8.10 Bosons and fermions in a quantum box. Physical systems like to have the lowest possible energy. For bosons, this can be achieved by putting all the bosons in the same, ground state energy level. Light photons behave like bosons. Fermions, on the other hand, are particles like electrons that must obey the Pauli exclusion principle. Thus, each energy level can be occupied by at most two fermions, corresponding to the two (spin up and spin down) states of an electron.

In 1924 Bose was a young Bengali physicist who was virtually unknown in the scientific world. His sixth scientific paper concerned a new derivation of the famous formula with which Planck had introduced both the concept of photons and h, his famous quantum constant. There are many stories in physics about papers which later became famous, being first rejected for publication. Bose's paper was one of these, but he had the good luck, or foresight, to send a copy to Einstein. Bose asked Einstein if he could arrange for publication in a German journal 'if he thought the work had sufficient merit'. At the time, Einstein was deeply engrossed in his search for a 'unified theory' of all the forces of Nature, but Bose's letter made him deviate temporarily from this, his main line of research. Einstein personally translated Bose's paper into German and sent it off to the journal with a note saying he believed that Bose's work constituted an 'important advance'. Over the course of the next few months, Einstein published several papers extending and clarifying Bose's work. In particular, it was Einstein who first noted the possibility of Bose particles, now called bosons, all 'condensing' down into the lowest energy state. We can see what this means by returning to the quantum problem of particles in a box. If, instead of electrons, we put in photons, then the lowest energy state is achieved by all of them occupying the lowest energy level. At normal temperatures, however, enough energy can be transferred in an ordinary collision for most of the bosons to be in excited states. Nonetheless, if we lower the temperature Einstein pointed out that 'from a certain temperature on, the molecules "condense" without attractive forces'. He went on to say: 'The theory is pretty, but is there also some truth to it?' This was in December, 1924; Bose had written to him in June of that year.

Albert Einstein (1879–1955) at the height of his powers in 1916. He had just completed his general theory of relativity as well as his important work on the absorption and emission of light from atoms, which we discuss in this chapter. Einstein received the Nobel Prize in 1921 for his work on the photo-electric effect – another vital contribution to quantum mechanics. In spite of the large part Einstein played in the origin of quantum theory, he remained unhappy with the conventional interpretation of the theory as championed by Heisenberg and Bohr. This is not to say he disputed that quantum mechanics worked, but, rather, that the theory in its present form, with uncertainty playing an essential role, was incomplete. In a letter to Born, who first introduced the probability interpretation of Schrödinger's waves, Einstein made his famous remark that God 'does not play dice'.

Einstein's proposed condensation of bosons ('Bose condensation') had, at first, the reputation of having 'a purely imaginary character' and not representing real observable physics. It was not until 1938 that Fritz London proposed that some strange effects observed with liquid helium could be understood in terms of a Bose condensation of helium atoms. Before we go on to describe these bizarre properties of helium we must first answer a more fundamental question. As we have said, matter-like particles such as electrons, protons and neutrons are all fermions, so why should helium be considered a boson? The reason is that the usual ^{4}He atom contains an even number of fermions: two protons and two neutrons in the nucleus, and two atomic electrons. Experiment tells us that elements with an even number of fermions can behave like bosons. Thus, liquid ^{4}He can undergo a Bose condensation at low temperatures and show remarkable 'superfluid' behaviour. On the other hand, elements with an odd number of fermions are found to obey the Pauli principle and act like fermions. Thus, liquid ^{3}He, with only one neutron in its nucleus, is a fermion and does not undergo a similar condensation to ^{4}He and has very different properties at low temperatures, despite its chemical similarity!

Helium has the lowest boiling point of any gas and was the last to be liquefied. In low temperature physics temperatures are usually given in 'kelvin' (symbol K), rather than in degrees Celsius or centigrade. Absolute zero is defined to be the zero of the kelvin scale and corresponds to about -273°C: there can be no temperatures lower than absolute zero. Towards the end of the 19th century, physicists in Paris, London and Cracow were vying with each other to produce the lowest temperature. For a long time it had seemed that liquefying hydrogen would be the last step on the road to absolute zero. Sir James Dewar announced the first hydrogen liquefaction to the Royal Society in London in 1898. In his experiments he had reached about 12 K. By this time, however, the rare helium gas had been discovered and it had become clear that helium liquefaction was the real goal. In 1904, Dewar estimated the temperature required to be around 6 K, but it was not

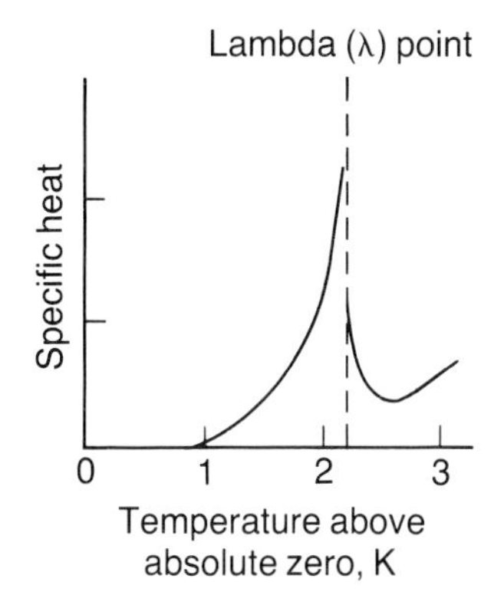

Fig. 8.11 The lambda point for liquid helium. This is a peculiar change that occurs at about 2.2 K above absolute zero and is evident in this 'lambda-like' curve formed by these measurements of the 'specific heat' of helium.

until 1908 that the Dutch physicist Kamerlingh Onnes, in Leiden, finally succeeded in liquefying helium. The boiling point of helium was in fact found to be about 4 K.

Liquid helium is now known to have many remarkable properties. It remains liquid even if cooled as close as possible to absolute zero. This is because of the large 'zero-point' motion of the helium atoms – the necessary quantum jiggling required to satisfy Heisenberg's uncertainty principle. Furthermore, at about 2 K, a dramatic change occurs. All boiling ceases and the liquid becomes very still. Moreover, other properties also change abruptly. Fig. 8.11 shows the variation with temperature of the 'specific heat': the amount of heat needed to raise the temperature of one gram of helium by one kelvin. Because the shape of this curve looks like the Greek letter lambda λ, this temperature is known as the 'lambda point'. Below the lambda point, the viscosity, or 'treacliness', of helium suddenly drops by a factor of about one million. Perhaps most surprising of all is the ability, below the lambda point, of liquid helium to 'creep' as a thin film along the walls of its container. If a beaker is lowered into a larger container of liquid helium, a thin film of helium forms over the entire surface of the beaker. This film then acts like a siphon through which helium can flow with almost no viscosity. Thus, no matter what the original difference in levels of the helium inside and outside of the beaker, the helium flows till they are equal! Kurt Mendelssohn recalled the discovery of this 'film transfer' phenomenon in the Clarendon Laboratory in Oxford, UK, with the following words:

'If the beaker is withdrawn from the bath, the level will drop until it has reached the level of the bath. If the beaker is pulled out completely, the level will still drop, and one can see little drops of helium forming at the bottom of the beaker and falling back into the bath. This is the sort of thing that makes one look twice and rub his eyes and wonder whether it is quite true. I remember well the night when we first observed this film transfer. It was well after dinner, and we looked around the building and finally found two nuclear physicists still at work. When they, too, saw the drops, we were happier'.

Fig. 8.12 Liquid helium below the lambda point is a 'superfluid' and shows some remarkable properties. This photograph shows how liquid helium can 'creep' up the sides of a container, flow over the top, down the outside and collect in drips at the bottom.

Kamerlingh Onnes (1853–1926) in his cryogenic laboratory in Leiden, The Netherlands. Onnes was the first to liquefy helium, and he received the 1913 Nobel Prize for this achievement. He also was the first to observe the phenomenon of superconductivity – the vanishing of the electrical resistance of some metals at very low temperatures.

All these peculiar properties of liquid helium are the result of the helium atoms 'condensing' into the lowest energy state, forming a quantum 'superfluid'. Since essentially all the atoms are in the same quantum state, they behave in a co-operative way and this is what gives the superfluid its unusual properties. As Feynman says at the beginning of this chapter, this is a striking example of the peculiarities of quantum mechanics being observed on a large scale. Without quantum mechanics, de Broglie, Heisenberg, Schrödinger and all, we would have no explanation for all these strange effects!

There is a short postscript to the story of liquid helium which serves as a good lead-in to our discussion of superconductivity. As we have said, liquid ^{3}He is expected to behave quite differently from liquid ^{4}He, since its atoms behave like fermions and cannot undergo Bose condensation. Experiment bears out this expectation, but a new form of Bose condensation is seen at a very much lower temperature, about 0.002 K! At this temperature, the weak attractive forces between two ^{3}He atoms are sufficiently strong for the pair to bind together and act like a boson. These pairs of ^{3}He atoms can then undergo a similar Bose condensation to that of the individual ^{4}He atoms. As we shall see, a similar mechanism is responsible for superconductivity.

Superconductivity

Soon after the discovery of the electron it was realized that many features of the ability of metals to conduct electricity could be explained in terms of the motion of electrons. The resistance to current flow is caused by electrons being 'scattered' by collisions with defects in the crystal lattice of the metal and by interaction with the vibrations of the crystal atoms. As the temperature is decreased, therefore, the atoms will vibrate less and less and

Fig. 8.13 This spectacular photograph of the 'fountain effect' is another example of the strange behaviour of liquid helium.

Fig. 8.14 Superconducting levitation. A small magnet floats over a superconducting dish. Superconducting currents flowing in the dish generate forces which repel the magnet and counteract the force of gravity.

one would expect the resistance of the metal to tend to a constant value. This is indeed what happens for many metals. It is therefore all the more surprising to find that the electrical resistance of certain metals falls suddenly to zero when they are cooled below a certain critical temperature. The electrical resistance of normal metals causes energy loss and heating; in these metals, by contrast, currents can be set up that persist for years. Such metals are genuinely 'superconductors'.

The phenomenon of superconductivity was discovered by Kamerlingh Onnes, 'the gentleman of absolute zero', at his laboratory in Leiden in 1911. Fig. 8.15 shows a graph of the resistance of mercury taken from his original paper. In 1933, another fascinating property of superconductors was discovered. If a magnetic field is applied to a superconductor, electrical currents are set up in the metal which conspire to cancel exactly the applied magnetic field. This exact cancellation is only possible because electric currents experience no resistance inside the superconductor. This leads to some very striking effects. A small magnet placed over a superconducting dish will float there because of the currents caused by the magnet in the dish. 'Superconducting levitation' has been seriously considered as a method for providing very smooth support for high speed trains.

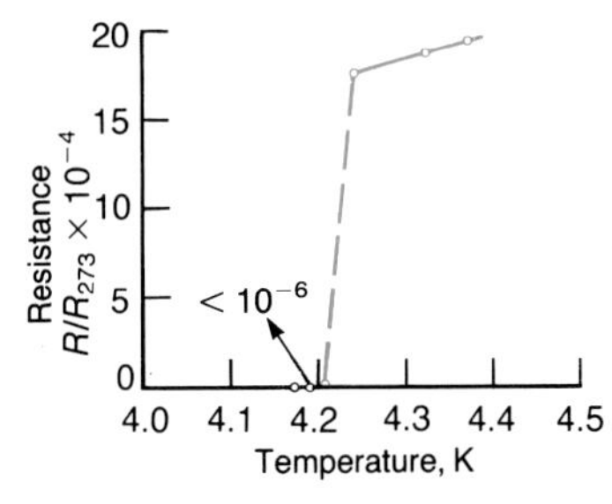

Fig. 8.15 Kamerlingh Onnes' discovery of superconductivity in 1911. A dramatic graph showing how the electrical resistance of mercury suddenly vanishes as the temperature is lowered below about 4.2 K above absolute zero.

How can superconductivity be understood? As early as 1935, in Oxford, the brothers Heinz and Fritz London – who did a lot of the early experimental and theoretical work on superconductors – realized that quantum mechanics must be an essential ingredient in any understanding of the effects. However, it was not until 1956 that Leon Cooper came up with the key observation. He showed that although two electrons normally repel each other, because of their electric charges, in a metal there is also an attractive force between them, caused by the attraction of the positively charged crystal lattice ions. Roughly speaking, an electron sitting between two positive ions in the lattice brings these ions a little closer together than normal, and another electron will therefore feel a small net attraction. There is therefore the possibility that two electrons will be bound together to form a 'Cooper pair'. These pairs are rather curious, however, in that they consist of

The BCS in the BCS theory of superconductivity – John Bardeen, Leon Cooper and John Schrieffer. The award of the 1972 Nobel Prize for their work made Bardeen the only person to win two Nobel Prizes in the same subject (with Brattain and Shockley he had discovered the transistor). Bardeen was a student of another famous quantum physicist, Eugene Wigner, who himself won the Nobel Prize in 1963. There are many examples of such 'father–son' Nobel Prize relationships.

electrons with opposite velocities adding up to a net momentum of zero for the pair. Moreover, according to Heisenberg's uncertainty principle, since the momentum of the pair is well defined, the pair of electrons must be very spread out in space. Each pair therefore occupies a region which is several thousand times larger than the size of the individual atoms and this same space is occupied by millions of other overlapping pairs.

Given our discussion of Bose condensation for ^{3}He, it is not too difficult to guess the form of the next step. The Cooper pairs act like bosons, and condense to form the superconducting state. While this is easy to say, it nevertheless proved difficult to come up with a quantitative and predictive theory of this condensation. This final step was taken in 1956 by a trio of physicists now universally known as 'BCS': John Bardeen, Leon Cooper and John Schrieffer. They were working at the University of Illinois, and, because of shortage of space, Bardeen and Cooper were sharing an office. Schrieffer was Bardeen's PhD student and had a desk with other theoretical physics students in a neighbouring building. They were trying to extend Cooper's idea of the formation of a single bound pair of electrons to all the electrons in the metal. Schrieffer later described what they were trying to do as a search for 'a quantum wavefunction to choreograph a dance for more than a million million million couples'. The problem seemed so hard that Schrieffer was thinking of changing his thesis problem to one on magnetism. At this crucial time, Bardeen had to go to Stockholm to receive his share of the Nobel Prize for the invention of the transistor, and before he left he urged Schrieffer to work on the problem for another month. In that month Schrieffer guessed at a manageable form for the wavefunction of the Cooper pair Bose 'condensate'. In the month that followed, B, C and S were able to show that their theory explained all the experimental data. Paradoxically, one finds that metals that are good conductors of electricity at normal temperatures have very little electron ion interaction and so will not be superconductors at low temperature. Rather, it is the bad conductors at normal temperatures that end up as superconductors.

There are many applications of superconductivity. Superconducting electromagnets are now used to obtain high magnetic fields without the usual power losses of electromagnets constructed using ordinary conductors for the windings in the coil. A problem arises in trying to go to very high magnetic fields. A magnetic field tends to be induced in the windings of the magnet itself, and too high a magnetic field can destroy the superconductivity of the coil. This problem can be alleviated by using so-called 'Type-II' superconductors. These are superconductors in which magnetic fields are not excluded completely from the metal but are able to penetrate the superconductor in thin 'tubes'. Very high magnetic fields can be produced by electromagnets using such superconducting wire. The property of superconductors in screening out magnetic fields can also be used to make improved electron microscopes.

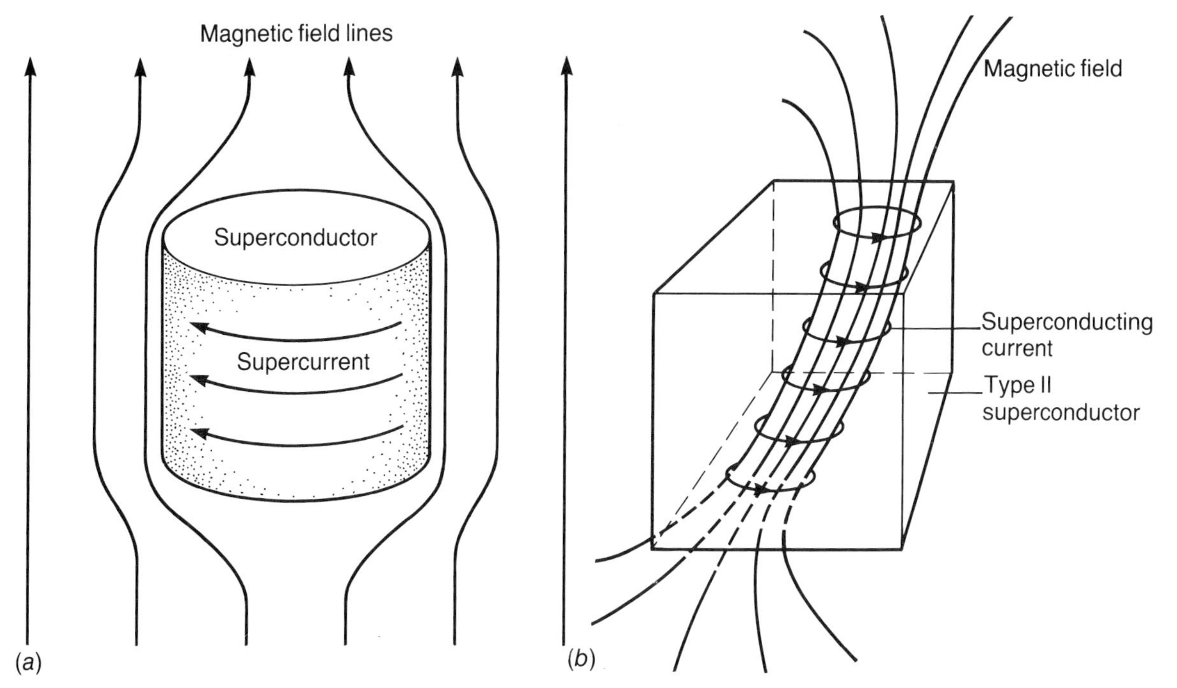

Fig. 8.16 Magnetic fields are excluded from superconductors. (a) In a type I superconductor, such as lead or tin, magnetic fields are expelled completely from the metal by the circulating supercurrents induced by the field. (b) By contrast, in a type II superconductor, the magnetic field can penetrate the metal in thin tubes.

Brian Josephson was only 20 when he made the discovery which was to win him a share of the 1973 Nobel Prize. Josephson was a student attending a course of lectures given in Cambridge, UK, by another great physicist and Nobel Prize winner, Philip Anderson. After class one day Josephson showed Anderson his calculations of tunnelling by superconducting Cooper pairs of electrons. Josephson's ideas have opened the door to superconducting interferometry, which has many applications in physics and technology.

Perhaps the best-known applications of superconductors are to the 'Josephson junction' and a device called a 'SQUID' – an acronym for 'superconducting quantum interference device'. Both use a discovery of a British PhD student, Brian Josephson. Philip Anderson, himself a Nobel prize winner, recalls giving a lecture course on solid-state physics in Cambridge, UK, in 1962, and having Josephson in the audience: 'This was a disconcerting experience for a lecturer, I can assure you, because everything had to be right or he would come up and explain it to me after class'. Josephson was studying the quantum theory of a superconductor–insulator–superconductor sandwich, in which the filling, the insulator, is only a very thin film. He showed that Cooper pairs were able to tunnel through the junction and give rise to some very interesting effects. One prediction was that a current would flow even if there was no voltage applied to the junction! He also worked out what would happen in the presence of a magnetic field and also in the presence of a very high frequency oscillating voltage together with a constant voltage. This last arrangement permits the most accurate measurement of the ratio of fundamental constants h/e (Planck's constant divided by the charge of the electron). The Josephson effect has also been used to measure incredibly small voltage differences and also as a sensitive radiation detector. By putting one or more Josephson junctions together in an electrical circuit, it is possible to make a device for measuring magnetic fields extremely accurately. These devices are the aforementioned SQUIDS, and they are now being used in such diverse fields as medicine and geology. All these applications are possible because the Bose condensation of the Cooper pairs in a superconductor allows us to observe quantum effects in the large instead of being restricted to atomic dimensions.

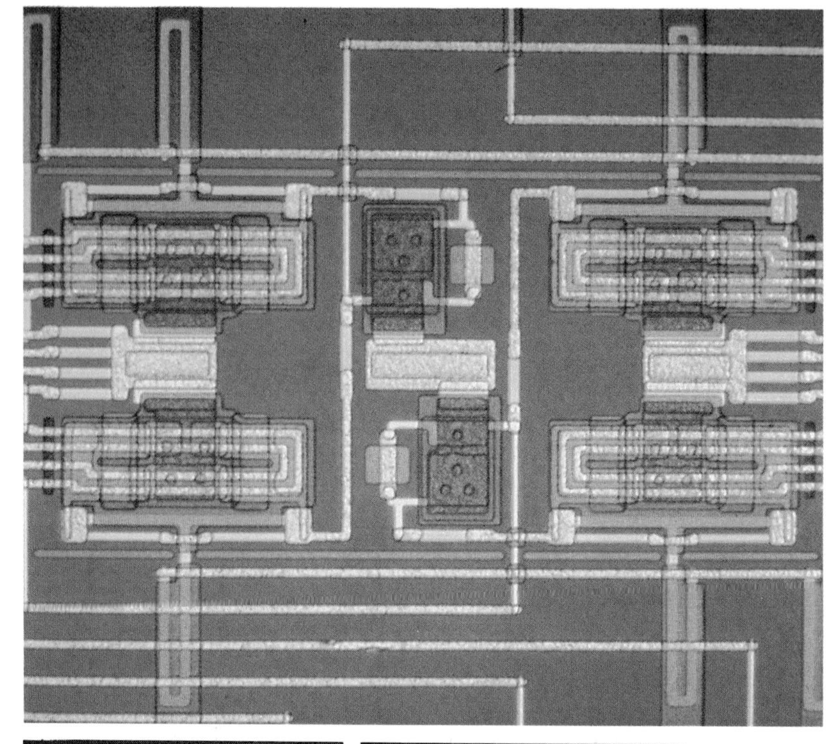

Fig. 8.17 Computers of the future may contain superconducting circuits with Josephson junctions acting as very fast switches. In this IBM chip the Josephson junctions lie underneath the four small brown circles in the brown regions. At present, it is unclear whether this superconducting technology will be commercially successful.

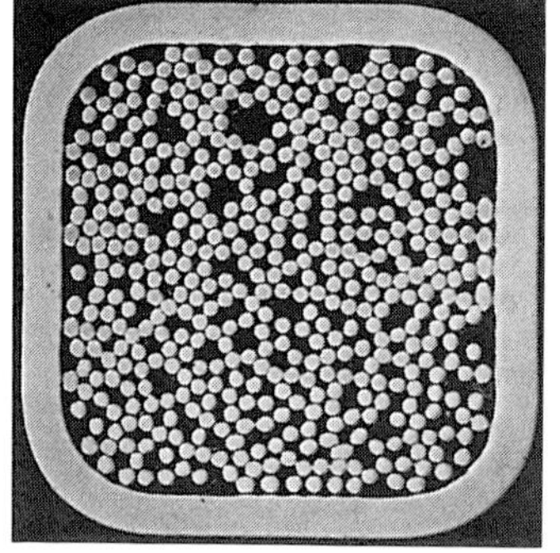

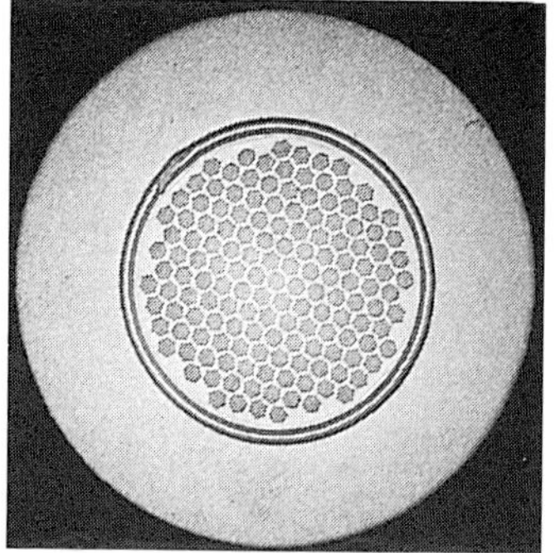

Fig. 8.18 A montage of photographs of superconducting cable designed to carry very large electric currents. The photograph at the upper left shows a steel pipe containing hundreds of superconducting wires. Liquid helium is pumped through the cable to maintain the low temperatures necessary for superconductivity to exist in these wires. At the upper right a magnified photograph of a single wire shows how thousands of superconducting filaments are arranged in hexagonal groups, all encased in a jacket of copper. The bottom photograph shows an electron microscope picture of a single group of these filaments.

9

Feynman rules

It is as though a bombardier flying low over a road suddenly sees three roads and it is only when two of them come together and disappear again that he realizes he has simply passed over a long switchback in a single road.

Richard Feynman

Dirac and antiparticles

We have seen in the earlier chapters that quantum mechanics, despite its inherent probabilistic nature, is capable of making successful predictions for an enormous range of phenomena. In the microscopic domain, there is no doubt that Newton's laws of classical mechanics must give way to quantum theory. There is another area where Newton's laws have been shown to be in need of modification, namely when the velocities of objects are close to the speed of light. Since light travels at approximately 300 000 km/s, the effects of these modifications, like those of quantum mechanics, are not usually apparent in everyday life. According to Einstein's theory of special relativity the energy, E, and momentum, p, of a particle are related by the equation

$$E^2 = p^2c^2 + m^2c^4$$

where c is the velocity of light and m is the mass of the particle when it is at rest. The more familiar relation between energy and momentum

$$E = p^2/2m$$

may be derived from the relativistic equation as an approximation valid when the velocity of the particle is much less than the velocity of light. It is also customary not to include the 'rest mass energy', mc^2, in this 'non-relativistic' expression for the energy. Einstein, Technical Expert Third Class at the Patent Office in Berne, Switzerland, had developed the special theory of relativity in 1905. The theory was well-understood and accepted by the physics community by the 1920s, when quantum mechanics was being invented. In particular, it was natural for Schrödinger to try to develop quantum mechanics starting from the relativistic energy–momentum equation above. After trying unsuccessfully to develop a relativistic wave equation that agreed with experiment, Schrödinger resorted to the approximate non-relativistic relation and found his famous equation, which was published in January, 1926. In spite of the great success of the Schrödinger equation, it was clear that this version of quantum mechanics was not valid for high speed electrons. Furthermore, the 'spin' angular momentum of the electron had to be tacked on to the theory in a rather unsatisfactory way. A relativistic equation was needed.

Albert Einstein worked at the Patent Office in Berne, Switzerland. After a crucial conversation with his friend and colleague at the Patent Office, Michele Besso, Einstein realized that a radical rethinking of the nature of time was needed and this led to his special theory of relativity. Nevertheless, Paul Dirac has said that if Einstein had not published his theory in 1905, someone else would have done so, soon after. Dirac also went on to say that without Einstein we would probably still be waiting for the general theory of relativity. About Dirac, Einstein once said: 'I have trouble with Dirac. This balancing on the dizzying path between genius and madness is awful'.

Paul Dirac was born in Bristol, UK, in 1902 and obtained a BSc degree in electrical engineering from the University of Bristol in 1921. Twelve years later he shared the Nobel Prize with Schrödinger 'for their discovery of new and productive forms of atomic theory', and had successfully predicted the existence of antimatter. Dirac was an extremely original thinker but notoriously reserved. Heisenberg told an amusing story about Dirac which illustrates both these qualities. The two of them were travelling to Japan from the USA by boat and Heisenberg liked to join in the social activities that went on in the evenings. At a dance once night, Heisenberg was enjoying himself dancing and Dirac, as usual, was sitting watching. As Heisenberg came back to his chair after a dance Dirac asked him 'Why do you dance?' Heisenberg replied 'Well, when there are some nice girls it is a pleasure to

Paul Dirac and Werner Heisenberg in 1933. Dirac's father was Swiss but he emigrated to England and became a language teacher in Bristol. Dirac was brought up to be bilingual in French and English, but remained extremely reserved in both languages. He was married to Wigner's sister and he shared the 1933 Nobel Prize with Schrödinger.

dance'. Dirac thought about this for a while; then, after about five minutes, he said 'Heisenberg, how do you know beforehand that the girls are nice?'

It is nonetheless rather curious that Dirac is still relatively unknown to the general public. He certainly was one of the 20th century's greatest physicists and his achievements are on a par with such great figures as Newton, Maxwell and Einstein. What did Dirac actually do? As Feynman says, 'Dirac got his answers by . . . guessing an equation'. The Dirac equation looks deceptively simple when written in the usual, very compact mathematical notation (see box below). For a relativistic version of quantum mechanics to ensue, the solutions of this equation must satisfy the correct relativistic relation between E and p. But, for a given momentum p, there are two possible solutions for the energy, namely

$$E = \pm \sqrt{(p^2c^2 + m^2c^4)}$$

One solution has positive energy as we would expect, but the second solution appears to have negative energy! How can negative energy solutions have any physical sense? Dirac's great achievement was to take them seriously and turn these apparently unwelcome solutions into a triumph of theoretical physics. His ingenious suggestion was that the negative energy levels did exist but were normally already occupied by electrons. Then, because of Pauli's exclusion principle, no ordinary positive energy electron can make a transition to any of these levels. Thus, according to Dirac, a quantum box, which is apparently 'empty', containing no positive energy electrons, in fact has a fully occupied 'sea' of negative energy electrons! This is not as ridiculous as it first sounds since if we now put some positive energy electrons into this box, we only measure the charge and energy relative to the 'empty' box. Thus, all the negative charge and negative energy of the empty box is unobservable. So far, the reader would have some justification for thinking this is all some wild theoretical fantasy, but, like any good theory, Dirac's picture of the 'vacuum' – as the empty box state is often called – makes some dramatic predictions. We know that if we shine light on an atom, electrons can absorb energy from a light photon and jump to an excited state. What happens if we shine light on an 'empty' box? According to Dirac, we should be able to excite one of these negative energy electrons up to a positive energy state. Instead of an empty box, we would then have a positive energy electron, together with a 'hole' in the Dirac sea. Relative to the normal empty box state, a box with a hole in the sea is lacking some negative energy and negative charge. Compared with the 'empty' vacuum state, therefore, a hole in the sea has positive energy and positive charge! The physical process we have described using Dirac's negative energy sea is the creation of an 'electron–positron pair' by a photon. The positron is the 'antiparticle' of the electron, a particle with the same mass but opposite charge.

$$E\psi = (-i\alpha \cdot \nabla + \beta m)\psi$$

Dirac equation for relativistic electron

As always in physics, a theory is only as good as its predictions. The positron was duly found in cosmic ray experiments by Carl Anderson in 1932, four years after Dirac wrote down his equation. The antiproton – the antiparticle of the proton – was discovered at Berkeley, California, in 1955.

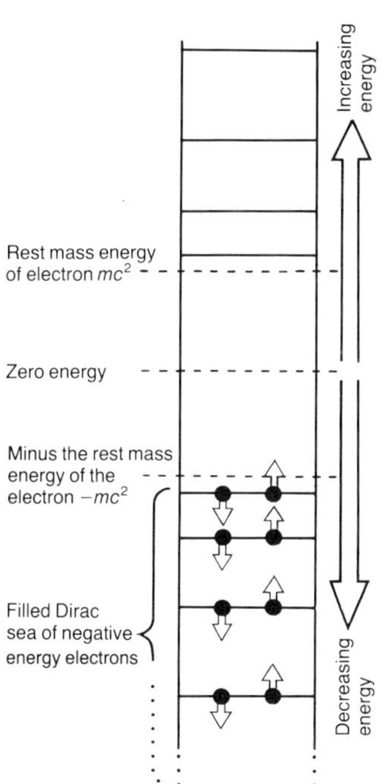

Fig. 9.1 Dirac's picture of the 'vacuum'. Solving the Dirac equation for relativistic electrons in a box leads to both positive and negative energy levels. Dirac made sense of this by supposing that in an empty box – the vacuum state – all the negative energy levels were filled. Then, according to the Pauli exclusion principle, if a positive energy electron is put into the box it cannot lose energy by falling into the negative energy levels.

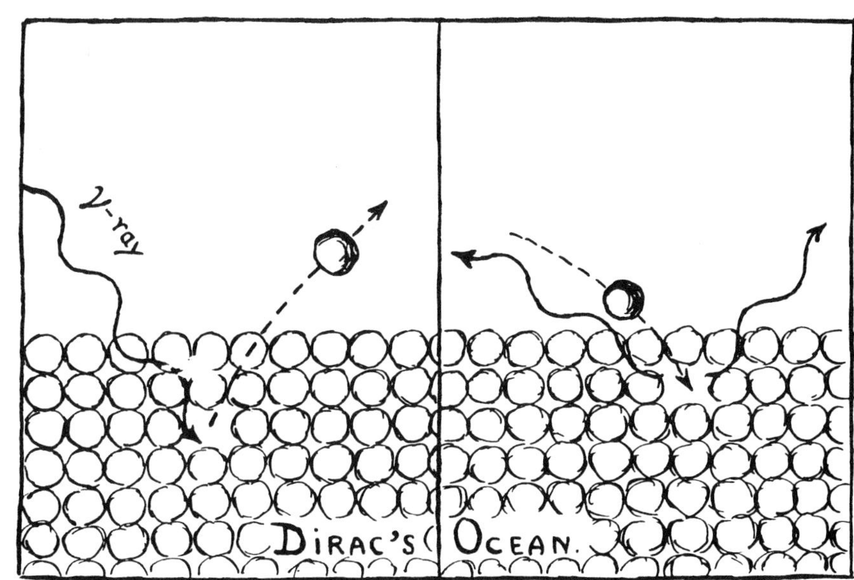

Fig. 9.2 Dirac's sea of negative energy electrons has observable consequences. In this picture from Mr Tompkins' adventures we see that a high energy photon or gamma ray can excite an electron from a negative energy level in the sea so that it appears as an ordinary positive energy electron. The 'hole' in the sea acts like a particle of positive charge and energy relative to the normal vacuum state. The photon has therefore created a particle–antiparticle pair. The second picture shows an electron jumping into a hole in the sea. This corresponds to the reverse process of annihilation of an electron with its positron antiparticle to produce high energy photons.

Its discovery had to wait for the construction of an accelerator which could provide enough energy for the creation of proton–antiproton pairs via the reaction

$$p + p \rightarrow p + p + p + \bar{p}$$

The converse process to pair creation was also predicted by Dirac. If we have a positive energy electron plus a positron 'hole' in our quantum box, then the electron can jump back into the sea and fill up the hole, leaving an empty box together with two photons to take away the 'annihilation' energy

$$e^+ + e^- \rightarrow \gamma + \gamma$$

Although the notion of Dirac's negative energy sea was the first way antimatter was predicted, it is a rather awkward and asymmetrical way of looking at antiparticles. The new feature of relativistic quantum mechanics, and one which is illustrated by both the pair creation and annihilation processes, is the possibility of transforming energy into matter. Thus, unlike the non-relativistic quantum mechanics of Schrödinger and Heisenberg, the number of quantum particles can change. Dirac's invocation of the filled negative energy sea of electrons, protons and so on, is just a cunning device to get round this problem. Moreover, since Dirac relies on the Pauli principle to prevent positive energy particles jumping to a lower energy state, this 'filled sea' trick will not work for bosons. As we shall see, there is a boson called a 'pi-plus' particle, π^+, which has a perfectly respectable 'pi-minus', π^-, as its antiparticle. The Dirac sea, with its apparently infinite negative

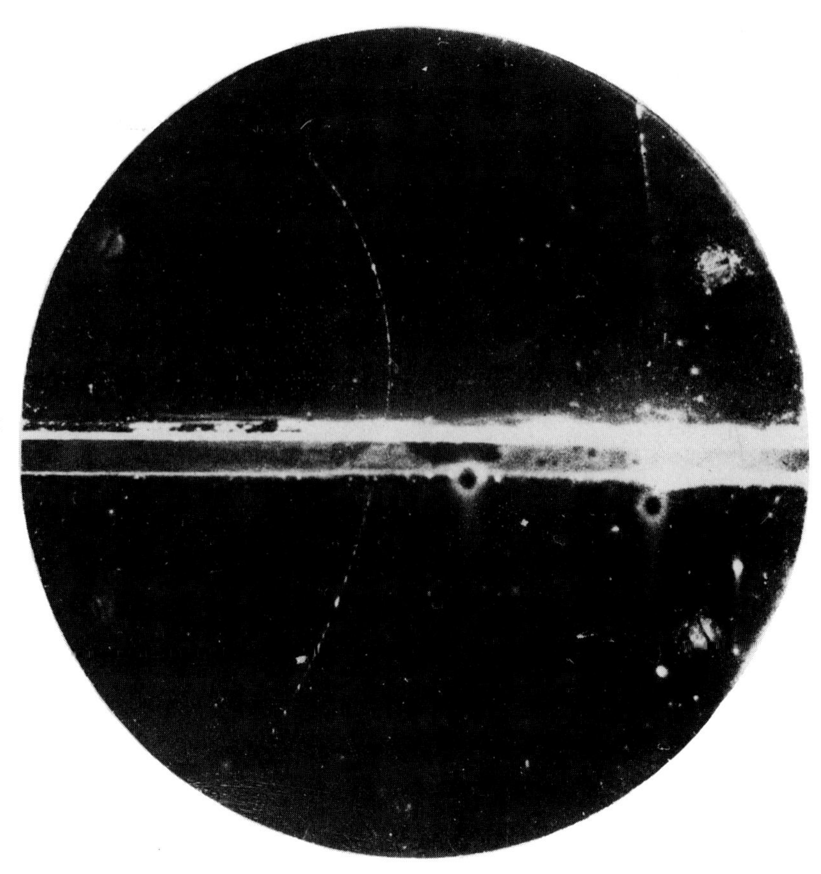

Fig. 9.3 A curved line of droplets shows the passage of a positron through a cloud chamber. The tracks are curved because there is a magnetic field throughout the chamber and the direction of curvature gives the sign of the electric charge. This photograph was taken by Carl Anderson and conclusively verified Dirac's prediction of antimatter. The lead plate in the middle of the chamber slowed down the positron as it passed through and resulted in the track above the plate being more tightly curved. This meant that Anderson knew that the track represented a positive charge going upwards rather than an ordinary electron going downwards. Nonetheless, Anderson had to take elaborate precautions to be sure that none of the Cal Tech undergraduates had played a joke on him by reversing the magnetic field.

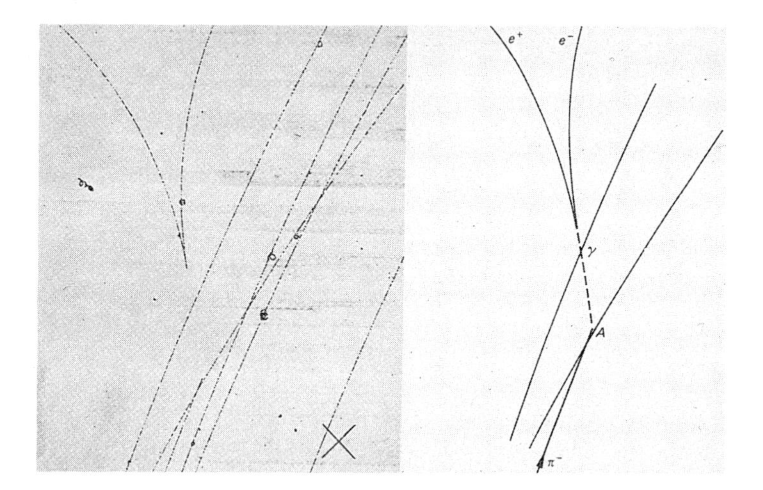

Fig. 9.4 A bubble chamber photograph of electron–positron pair creation. Only charged particles leave tracks in the chamber and the photon leaves no direct trace.

charge and mass, disappears from a proper 'many-body' quantum theory which allows for particle creation and annihilation processes right from the outset. Such many-body theories are examples of 'quantum field theories'. The relativistic quantum field theory describing the interactions of electrons and photons is known as 'quantum electrodynamics', or 'QED' for short. QED combines Maxwell's equations of electromagnetism, quantum mechanics and relativity. It is the most successful theory physicists have yet constructed and it has been tested to an astonishing accuracy. To demonstrate that this is no idle boast, consider the spin of the electron that caused Schrödinger so much trouble. The spinning electron acts like a little

Fig. 9.5 A picture of the Bevatron accelerator in Berkeley, California. This machine was the first accelerator to have enough energy to produce antiprotons.

magnet and the strength of the 'magnetic moment' of the electron can be calculated in QED. The result may be expressed in terms of the 'g-factor' of the electron. The classical value predicted for g is

$$g_{\text{classical}} = 1.0$$

which is less than half the value predicted by QED

$$g_{\text{quantum}} = 2.0023193048(8)$$

When this is compared with the experimentally measured value

$$g_{\text{experiment}} = 2.0023193048(4)$$

the agreement is indeed impressive. The numbers in brackets are the theoretical and experimental uncertainties in the last digit.

Unfortunately, any detailed description of quantum field theory involves a large amount of advanced mathematics that is far beyond the scope of this book. Fortunately for us, Feynman has provided a beautiful intuitive and pictorial approach to quantum field theory. This is the subject of the next section.

On the right of this photograph is Carl Anderson, who won the Nobel Prize for the discovery of the positron, while on the left is his student, Donald Glaser, who won the Nobel Prize for the invention of the bubble chamber. There is a bar in Ann Arbor, Michigan, where Glaser is said to have got the idea for the bubble chamber by watching the bubbles in a glass of beer.

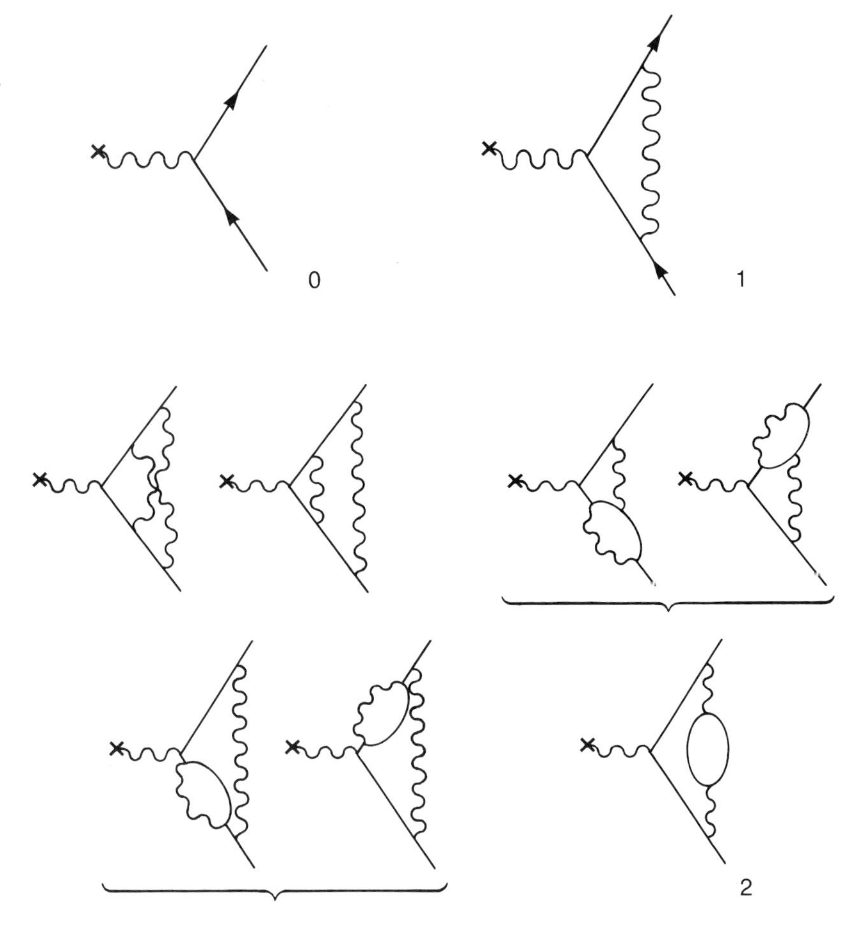

Fig. 9.6 Feynman diagrams needed for the calculation of the magnetic moment of the electron. The straight lines represent electrons and the wiggly lines photons. Diagrams with more and more photon lines are less and less important but each 'loop' diagram involves doing very complicated integrals.

Feynman diagrams and virtual particles

Feynman's way of looking at negative energy states takes a little getting used to, but in the end is very helpful. To see what is involved let us imagine a 'scattering' experiment involving electrons. Picture this experiment rather as if you were flipping a ball in a pin-ball machine and the ball makes two collisions on its way to the top of the table. In one try both the collisions are glancing collisions which merely deflect the ball a bit but do not stop it moving up the table. The second case shows a much more energetic collision in which the first collision sends the ball back down the table before it undergoes another violent collision and heads up the table again. In the case of a scattering experiment with electrons we can draw two similar trajectories for the electron, but with the difference that we now interpret the table as a 'space–time' plot, with the time axis going up the table. The first case looks entirely normal – the electron is just deflected by the collisions but continues on its way in the same general direction. The second case now looks very peculiar – the electron appears to have been scattered backwards in time! Feynman proposed that in relativistic quantum mechanics this bizarre possibility was allowed if the electron going 'backwards in time' had negative energy. The physical interpretation of this situation is again in terms of electron–positron pair creation. Absorbing negative energy and charge 'from the future' decreases the total energy and charge of the scattering centre. This has the same effect on the energy and charge of the centre as if positive energy and charge were emitted into the future. Thus,

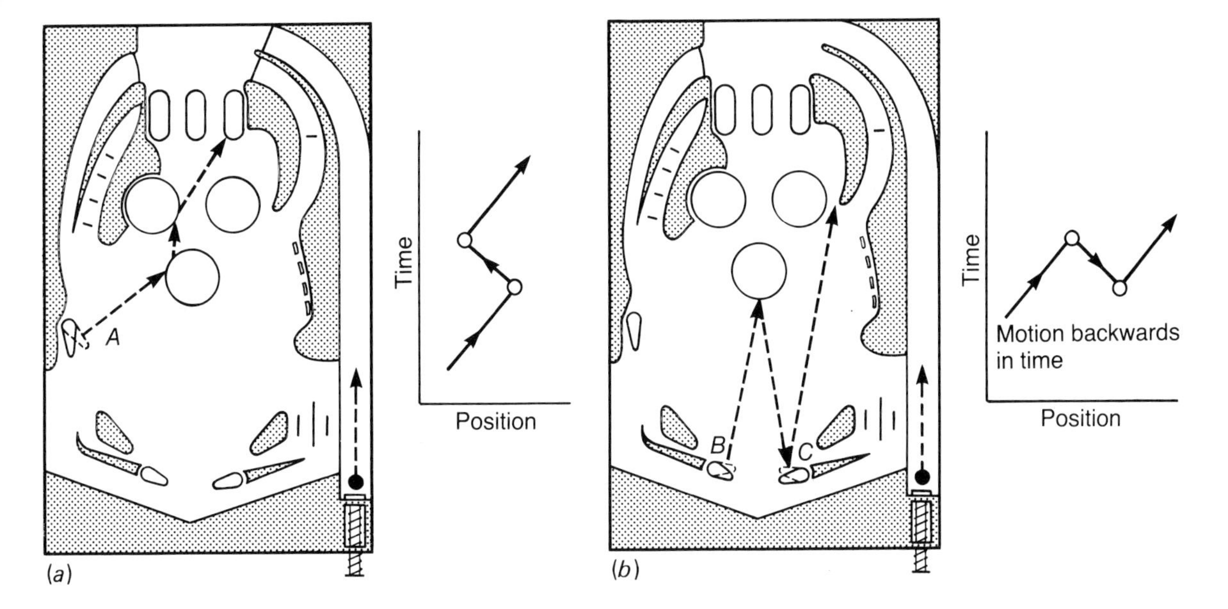

Fig. 9.7 Two possible paths of a ball in a pin-ball machine.
(a) The flipper has hit the ball so that it makes a glancing collision on its way up the table. Alongside this we show a similar path for the scattering of an electron. The axes of the graph are labelled with time t going upwards and space x across.
(b) In this case, the flipper has hit the ball so that it suffers a very hard collision and bounces back down the table before being hit back up again. In relativistic quantum mechanics there is a similar path for an electron on a 'space–time' graph. However, things now look very peculiar since the first scattering appears to cause the electron to be scattered 'backwards in time'!

this apparently absurd trajectory for the electron corresponds to the creation of an electron–positron pair at the first scattering centre. The positron then travels forwards in time to the second scattering centre where it is annihilated by the original incoming electron.

We can now see the relevance of Feynman's quotation at the beginning of this chapter, which is taken from his original paper entitled 'Space–time approach to quantum electrodynamics', published in the *Physical Review* in 1949. We have been able to avoid dealing explicitly with the many-particle nature of the pair creation process by treating the trajectory of the electron and positron as a single electron 'world line' in the same way that the road is actually a single road although it sometimes appears differently to the pilot. There is one great advantage of Feynman's interpretation of the relativistic negative energy states: using negative energy states moving 'backwards in time' to correspond to positive energy antiparticles moving forwards in time, works just as well for bosons as it does for fermions. Needless to say, we should stress that all this is only a device to get the right answer without having to use the complicated machinery of quantum field theory. Nothing, as far as we know, actually travels backwards in time!

There is one other key idea that we must introduce in our account of relativistic quantum mechanics. This is the idea of a virtual particle. In our discussion of tunnelling, back in chapter 5, we pointed out that one way of looking at tunnelling was in terms of the energy–time uncertainty relation

$$(\Delta E)\,(\Delta t) \approx h$$

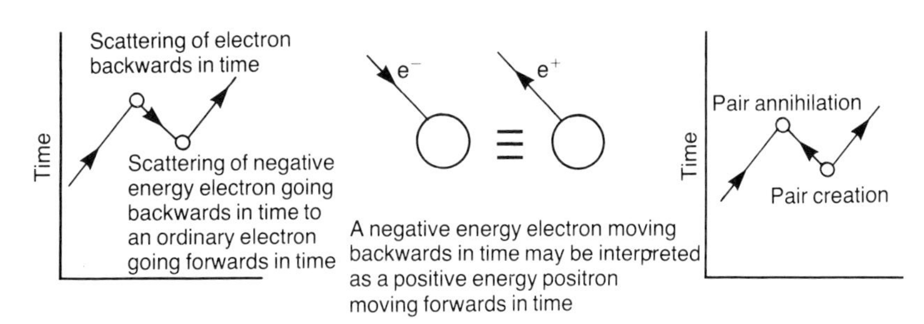

Fig. 9.8 Feynman realized that these 'backwards in time' lines should be understood as pair creation followed by pair annihilation. Negative energy electrons travelling backwards in time are equivalent to positive energy positrons travelling forwards in time.

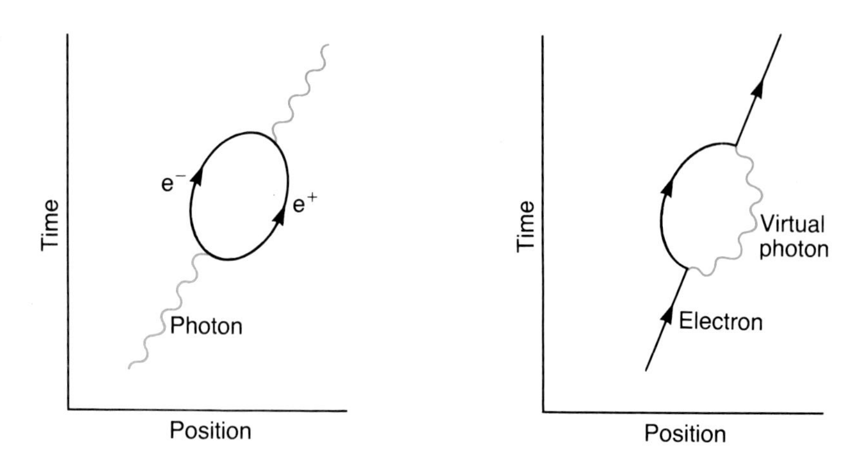

Fig. 9.9 Feynman diagrams for 'virtual' processes. According to the uncertainty principle energy can be 'borrowed' to allow particle creation provided it is 'repaid' in a short enough time.

In this context, this means that we can 'borrow' an energy ΔE and this loan will be undetected so long as we put it back within a time $\Delta t \approx h/\Delta E$. With the possibility of particle creation in relativistic quantum mechanics this means that a particle need not always remain the same particle. Enough energy could have been borrowed to create another particle or pair of particles which can only exist for a very short time. A photon, for example, may borrow enough energy to turn into a virtual electron–positron pair. These particles can exist for a very short time before they recombine back into a photon. Such processes are called 'virtual processes' and the particles created on the borrowed energy, 'virtual particles'. The probability amplitude for such 'virtual' processes may be calculated in QED, and Feynman invented a system of diagrams to help in evaluating such amplitudes. Similar virtual processes are possible for electrons. Such Feynman diagrams also help calculate the probability amplitudes for scattering processes. Consider electrons scattering from protons. In terms of Feynman diagrams, the scattering process may be visualized in terms of a virtual photon being 'exchanged' between one of the quark constituents inside the proton and the electron. The importance of Feynman diagrams is not just the fact that they give us a helpful way of visualizing all the possible virtual processes. Feynman also gave definite rules for calculating the amplitude of any diagram, no matter how complicated. Fig 9.6 shows some of the diagrams needed for the calculation of the electron magnetic moment that we mentioned earlier.

Feynman diagrams also have the nice property that the same diagram can be used for processes involving antiparticles. Fig. 9.10 shows the simplest diagram for electron–quark scattering by exchange of one virtual photon. In the diagram, the time axis is assumed to be pointing towards the top of the page. Now rotate the diagram, and remember that lines with arrows pointing backwards in time must be interpreted as antiparticles moving forwards in time. We see that the diagram now represents the physical process of electron–positron annihilation into one virtual photon which then creates a quark–antiquark pair.

Fig. 9.10 The process of electron–quark scattering is represented by the Feynman diagram shown in (a). The same diagram, rotated on its side, (b), is able to predict electron–position annihilation to create a quark–antiquark pair.

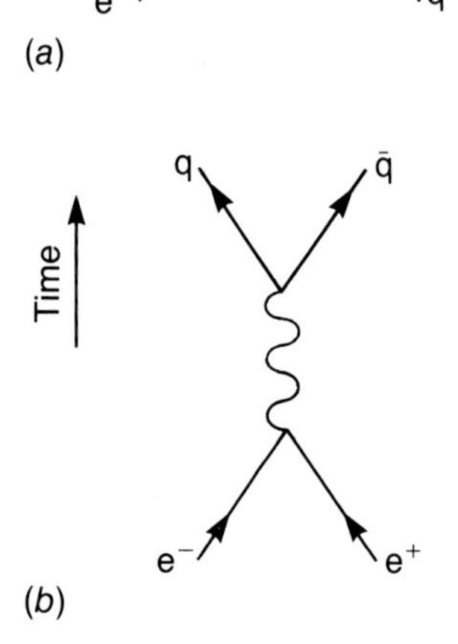

Zero-point motion and vacuum fluctuations

Dirac's picture of the vacuum seemed complicated enough, but we now seem to have replaced it with one that looks much more complicated. Instead of a place where nothing happens, the 'empty' box should now be regarded as a bubbling 'soup' of virtual particle–antiparticle pairs! For example, in our first discussion of an electron scattering backwards in time, it may have seemed as if the positron that was created in the first scattering

Hendrik Casimir was born in The Netherlands and was a student of Niels Bohr. He later became the research director of the Philips laboratories in Eindhoven.

needed to know in advance that it was going to be annihilated by the incoming electron in the later scattering. We now realize that the vacuum is full of such virtual electron–positron pairs, all existing for fleeting moments, and this was just one positron that got caught!

The 'vacuum' of relativistic quantum mechanics – or, strictly speaking, the 'ground state' of the relativisitc quantum field theory – has other interesting observable effects. When we apply quantum mechanics to the vibrations of atoms in a crystal, the vibrational waves set up in the crystal turn out to have particle-like aspects, just like photons. These quantum lattice vibrations are called 'phonons'. If the crystal lattice is cooled down so that no vibrational phonons are excited, there must still be some 'zero-point motion' of the atoms. It is this zero-point motion that prevents liquid helium from solidifying, as we discussed in the preceding chapter. What is the relevance of this crystal lattice to our discussion of the real physical vacuum? Well, in the same way that phonons are quantum objects associated with vibrations of the crystal positions, so too may we regard photons as associated with 'vibrations' of the electromagnetic field. Then we see that, as in the case of phonons, there must be a 'zero-point motion' for the electromagnetic fields. Remarkably enough, these 'vacuum fluctuations' of the electromagnetic field have some experimentally testable consequences.

Fig. 9.11 The core of the cluster of galaxies in the constellation of Corna Berenices. The very bright object is a star in our own galaxy, but nearly all the other objects in the picture are galaxies about 300 million light years away.

The most famous application of these ideas is the so-called 'Lamb shift' in the hydrogen spectrum. If one looks very carefully at the spectral lines of hydrogen one finds that there is a tiny splitting between the $n=2$, orbital angular momentum $L=1$ and $L=0$ energy levels that cannot be accounted for even by including the relativistic electron spin. The vacuum fluctuations of the electromagnetic field cause the electron in the hydrogen atom to jiggle about a little. The effect of this jiggling can be calculated and agrees remarkably well with Lamb's experimental measurement. The existence of these vacuum fluctuations also enabled Dirac to explain how excited electrons could undergo 'spontaneous emission' of a photon and jump to a lower state. The transition is in fact stimulated by the zero-point motion of the electromagnetic field.

There is another curious observable effect that originates with these vacuum fluctuations and which we may call a 'vacuum force'. In the physical vacuum, all possible wavelengths of fluctuations are allowed for the electromagnetic field. Consider a system consisting of two large metal plates set up face to face. Since the electromagnetic field is required to vanish on these plates, only those vibrations of the field for which this happens are allowed. Thus, since some of the normal vacuum wavelengths must now be absent, the zero-point energy of the electromagnetic field is altered. Detailed calculations show that this results in a tiny attractive force between the plates. This phenomenon is known as the 'Casimir effect', after the famous Dutch physicist who had been a student of Bohr. The existence of such a force was verified experimentally in 1958.

George Gamow is probably best remembered for his work on the Big Bang theory of cosmology. The picture shows Gamow emerging from a bottle of 'YLEM' – Gamow's word for the primordial neutron-dominated material out of which he believed the universe was born. Nowadays, physicists no longer believe that the matter in the early universe was mostly neutrons, but Gamow's idea of a hot and dense origin to the universe is now a cornerstone of modern cosmology.

Hawking radiation and black holes

Particle–antiparticle pair creation is now thought to be of relevance in the theory of black holes. In chapter 7 we encountered black holes as the final stage in the evolution of a very massive star. Once a black hole is born from a dying star it is easy to see how it might grow. Indeed many galaxies and the mysterious objects called 'quasars' – enormously powerful 'quasi-stellar radio sources' – are now thought to contain gigantic black holes. If black holes absorb all the radiation that arrives, how can such objects be seen? In fact, as stars and other interstellar matter are attracted towards the black hole, charged particles are accelerated and radiate electromagnetic energy which we can observe. However, once matter has been sucked inside the so-called Schwarzchild radius of the black hole, nothing, not even radiation, can overcome the gravitational forces and escape from the black hole.

The discussion above relates to black holes with masses ranging from three or four times the mass of our Sun to masses hundreds of millions of times greater than this in a quasar. It is much harder to see how very low mass black holes might be formed. Stephen Hawking, a cosmologist in Cambridge, UK, has suggested that black holes with a whole range of masses could be formed in the very early stages of the universe. To understand something of the conditions that most cosmologists believe prevailed soon after the creation of the universe, we must make a short detour and discuss some of the evidence for the expanding universe and the 'Big Bang' of creation.

The 'Big Bang' theory of cosmology was put forward by George Gamow and others to explain the observed expansion of the universe as we see it today. Fig. 9.11 shows the core of a large cluster of galaxies. Each galaxy contains a vast number of stars, as does our own Galaxy, the Milky Way. In the universe, we observe that clusters of galaxies are roughly evenly distributed throughout space. Fig. 9.12 shows galaxies in increasingly distant clusters, together with the measured spectra of light emitted from each galaxy. These spectra can be used to tell us how fast the galaxy is

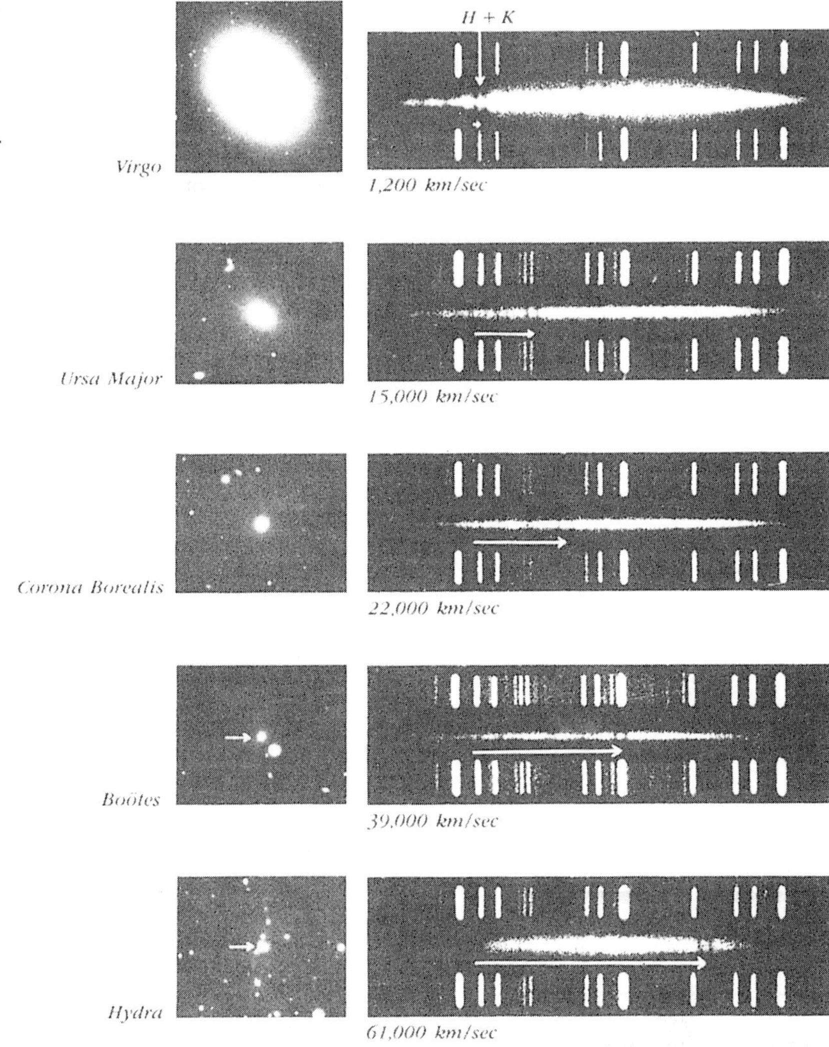

Fig. 9.12 Five galaxies together with their absorption line spectra. Compared with the reference spectra above and below, the absorption lines are seen to be further to the right for the further away galaxies. Since this 'Doppler' shift is believed to be due to the speed of the galaxy away from us, this shows that the further away a galaxy is from us, the more rapidly it is moving away. This is Hubble's law.

This painting by Wimmer shows Joseph Fraunhofer and his spectroscope. Fraunhofer discovered that the spectrum of the Sun was crossed by many dark lines. Fifty years later Kirchhoff was able to interpret these lines as absorption line spectra characteristic of the atoms in the Sun's atmosphere. On Fraunhofer's tombstone is the epitaph 'Approximavit sidera' *('He approached the stars').*

moving away from or towards us. Like the Sun, these galaxies emit light of all wavelengths – a 'continuous spectrum'. As light from the Sun travels through the outer layers of the Sun's atmosphere, some of the photons have just the right energy to excite electrons in the gas atoms making up these outer layers. Thus, photons with wavelengths characteristic of elements in this solar atmosphere will be missing from the light spectrum we see. This type of line spectrum is called an 'absorption' spectrum and such spectra can be used to identify what elements are present in the Sun. Fig. 9.12 shows similar absorption spectra from the various galaxies. What we find out from these absorption spectra is that the characteristic absorption lines of the various elements appear at longer wavelengths – they are 'red shifted'. This effect is similar to the familiar Doppler shift in the pitch of a train's whistle as it first comes towards and then travels away from you. These red shifts are interpreted as evidence that the galaxies are all moving away from us, and from each other.

The universe therefore appears to be expanding, and, moreover, the further away from us a galaxy is, the faster it appears to be moving away. This famous observation is known as Hubble's law:

$$v = H \times d$$

speed of recession = Hubble constant times distance away from us

Edwin Hubble (1889–1953) shown here with his cat, Nikolas Copernicus. He was born in Missouri and was a good enough athlete to consider a career in heavyweight boxing. After initially taking up law, Hubble returned to astronomy and said 'Even if I were second rate or third rate, it was astronomy that mattered'.

after the famous US astronomer Edwin Hubble. Incidentally, this observation does not put us at the centre of the universe. Imagine baking a currant loaf; as the dough rises all the currants see all the other currants moving further away from each other, and the further away a currant is, the faster it will be moving away.

This picture of an expanding universe carries with it the implication that at earlier times all the galaxies and matter must have been much closer together. This is the motivation for the Big Bang model of cosmology which extrapolates this expansion right the way back in time until all the matter was compressed together and enormously hot and dense. The huge energy densities that are present in the early stages of this Big Bang could have compressed matter so much that black holes with very small masses were created. These 'mini black holes' could have masses from as small as a few grammes up to the mass of a small planet. Hawking called these objects 'primordial' black holes.

Hawking also developed the theory of particle creation near such black holes. From our previous discussion of virtual particles, we know that the 'vacuum' can be regarded as a sort of bubbling 'soup' of virtual particle–antiparticle pairs. What Hawking proposed was that one member of such a pair could be captured by the black hole while the other member escaped into the surrounding space. How could this come about? Surprisingly, the key ingredient in the mechanism of such 'Hawking radiation' is an understanding of ordinary gravitational 'tidal forces'.

There is a short story called *Neutron Star*, written by Larry Niven, a science fiction writer popular amongst science students at Massachusetts and California Institutes of Technology. The background to the story is that the galaxy's main manufacturer of space-ship hulls, an alien race known as

Fig. 9.13 An optical picture of the quasar 3C273 over-exposed to bring out the faint jet. Quasars are believed to be due to matter falling into a large black hole.

Fig. 9.14 A currant loaf model for the expansion of the universe. As the loaf is baked the whole loaf expands and the currants move apart. Each currant sees all the others moving away.

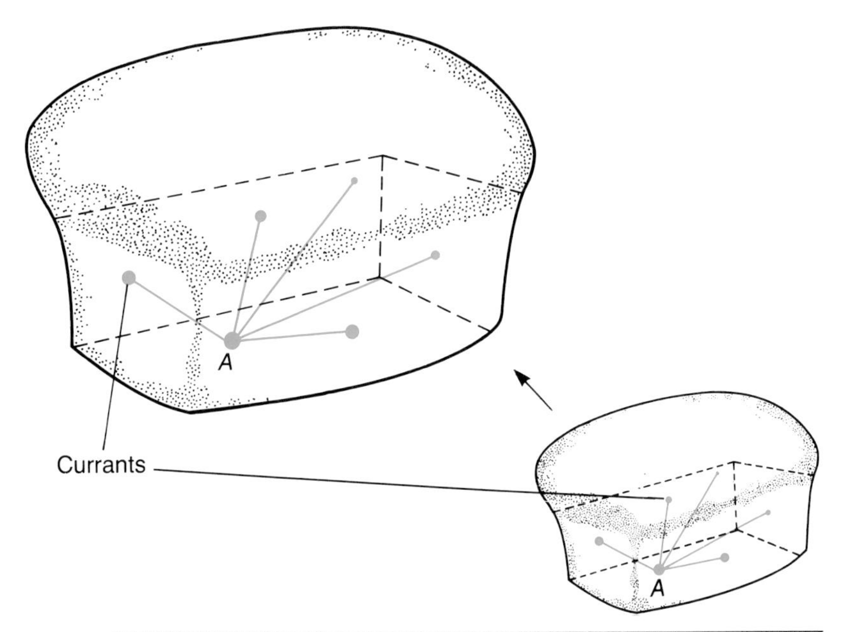

Fig. 9.15 The cover picture of Larry Niven's book Neutron Star. *The story first appeared in 1966 – one year before pulsars were discovered.*

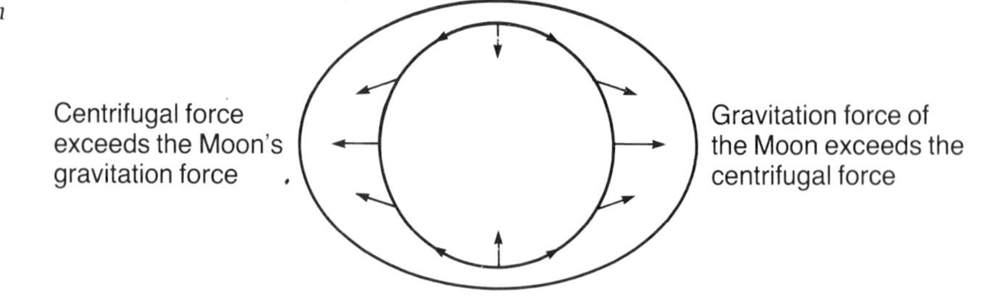

Fig. 9.16 A simplified diagram to illustrate the origin of the Earth's tides.

the puppeteers, are worried. Some unknown force has been able to penetrate their supposedly invulnerable 'No. 2 General Products hull' and kill the occupants who were on an exploratory voyage to a neutron star. The puppeteers, being inveterate cowards, blackmail the hero, Beowulf Schaeffer, into repeating the trip. Needless to say, the hero survives, and in the process realizes that the mysterious force is nothing but the familiar tidal force due to gravity. Since the puppeteers do not know what a tide is, Schaeffer is able to deduce that their secret home planet does not have a moon, and then blackmails them into paying him a fortune.

Stephen Hawking pictured with his son. Despite his appalling physical handicap Hawking has made remarkable contributions to astrophysics and cosmology.

What are these mysterious tidal forces and how could they kill or cause Hawking radiation? Fig. 9.16 shows an idealized picture of the Earth surrounded by a continuous ocean of water. On the side nearest the Moon, gravity attracts the water most strongly and is greater than the centrifugal force caused by the rotation of the Moon about the Earth. This therefore causes the water surface to bulge towards the Moon. On the other side of the Earth, further away from the Moon, the gravitational force on the water is smaller than the centrifugal force and the water surface bulges away from the Moon. The existence of these two bulges is the reason why we have two high tides for every daily rotation of the Earth. This is the effect of the Moon's gravitational attraction on the Earth. What is the effect of that of the Earth on the Moon? Since the Moon does not rotate relative to the Earth, a rock at the nearest point of the Moon and a rock at the point furthest away from the Earth are moving in two concentric orbits with the same orbital speed. If they were not part of the same Moon, the two rocks would naturally move in different orbits – since they experience different gravitational forces. Thus, the tidal force of the Earth on the Moon is trying to pull the Moon apart. Similarly, a spaceman in orbit round a neutron star or a black hole would experience huge gravitational tidal forces trying to tear him apart. It is these same tidal forces that can cause a particle–antiparticle pair to become separated if they are created in the enormous gravitational field of a primordial black hole. In this way it is possible for one of the pair to fall into the hole while the other escapes into the surrounding space. The black hole will thus appear to radiate particles.

This Hawking radiation by mini black holes corresponds to mass energy streaming away from the black hole into space. The mini black hole will thus 'evaporate' and the final stages of this process would be observable as a tremendous explosion of energy. Most primordial black holes produced in the Big Bang must have exploded long ago. However, if primordial black holes of all masses were created, some black holes should be exploding in our epoch. So far, none have been seen.

10

Weak photons and strong glue

Now we are in a position in physics that is different from any other time in history (it's always different!). We have a theory, . . . so why can't we test the theory right away to see if it's right or wrong? Because what we have to do is calculate the consequences of the theory to test it. This time, the difficulty is this first step.

Richard Feynman

The double-slit experiment revisited

In this chapter we turn to recent advances in the understanding of the fundamental forces of Nature. As we have said in earlier chapters, the combination of classical electromagnetism, quantum mechanics and relativity provides an astonishingly successful description of electromagnetic forces. The resulting theory is called quantum electrodynamics, or QED for short. For over 50 years physicists have been searching for similarly successful theories to describe not only the weak forces responsible for natural radioactivity, but also the strong forces that hold the nucleus together. It was not until the mid-1970s that any real progress was made, and this progress is the subject of this chapter.

Particle physicists now have a unified theory that combines both the electromagnetic and weak forces. This theory has recently been spectacularly verified by experiments at the CERN laboratory in Geneva. We shall describe the theory and these experiments in more detail below. However, not content with this *tour de force*, most physicists also believe that we have at last discovered the correct theory of the strong nuclear force. We now have a theory of the proton and neutron in terms of their quark constituents (see chapter 3). Why are we still unsure whether this theory, 'quantum chromodynamics', or 'QCD', is right? As Feynman explains in the introductory quotation to this chapter, the trouble is that it turns out to be extraordinarily difficult to calculate things with this theory. The basic problem is that the theory describes the forces in terms of quarks, but no quarks have ever been seen. The interactions between the quarks seem to be cunningly arranged so that we can only ever observe states of three quarks, like the proton and neutron, or states of a quark and an antiquark, like the mesons we shall meet in the next section. This non-observability of free quarks is called the 'quark confinement' problem, and how the theory solves this is not yet understood. We will discuss some speculative ideas about confinement below.

James Clerk Maxwell (1831—79) made original contributions to many areas of physics and was the first to suggest that Saturn's rings were composed of myriads of tiny particles. His most important work was putting Faraday's 'field ideas' into a precise mathematical form and unifying electricity and magnetism in one theory of electromagnetism. Maxwell's equations for electromagnetism were first published in 1865 and remain unchanged today, despite the development of both quantum mechanics and relativity. He died of cancer at a relatively young age and did not live to see Hertz verify his prediction of electromagnetic waves.

How is the double-slit experiment relevant to all this? Well it appears that Nature has been surprisingly kind to us. All the above theories have at their heart the same basic principle, and we can gain some understanding of this by looking again at the double-slit experiment with electrons that we first discussed in chapters 1 and 2. This principle is usually referred to by a rather inappropriate and intimidating name, namely 'gauge invariance', but we will try to convince you that the fundamental idea is really very simple and appealing.

Fig. 10.1 shows the double-slit experiment with electrons once again. As we discussed in chapter 1, the probability of arrival of electrons at the screen is predicted by assuming that electron waves from each slit overlap and interfere. Whether there will be lots of electrons arriving at any given point, or none at all, depends on whether the waves arriving from the two slits are both crests (in phase) or a crest arriving with a trough (out of phase). Suppose we now insert, between the slits and the screen, a thin sheet of material, as shown. In a similar fashion to the interference experiment with

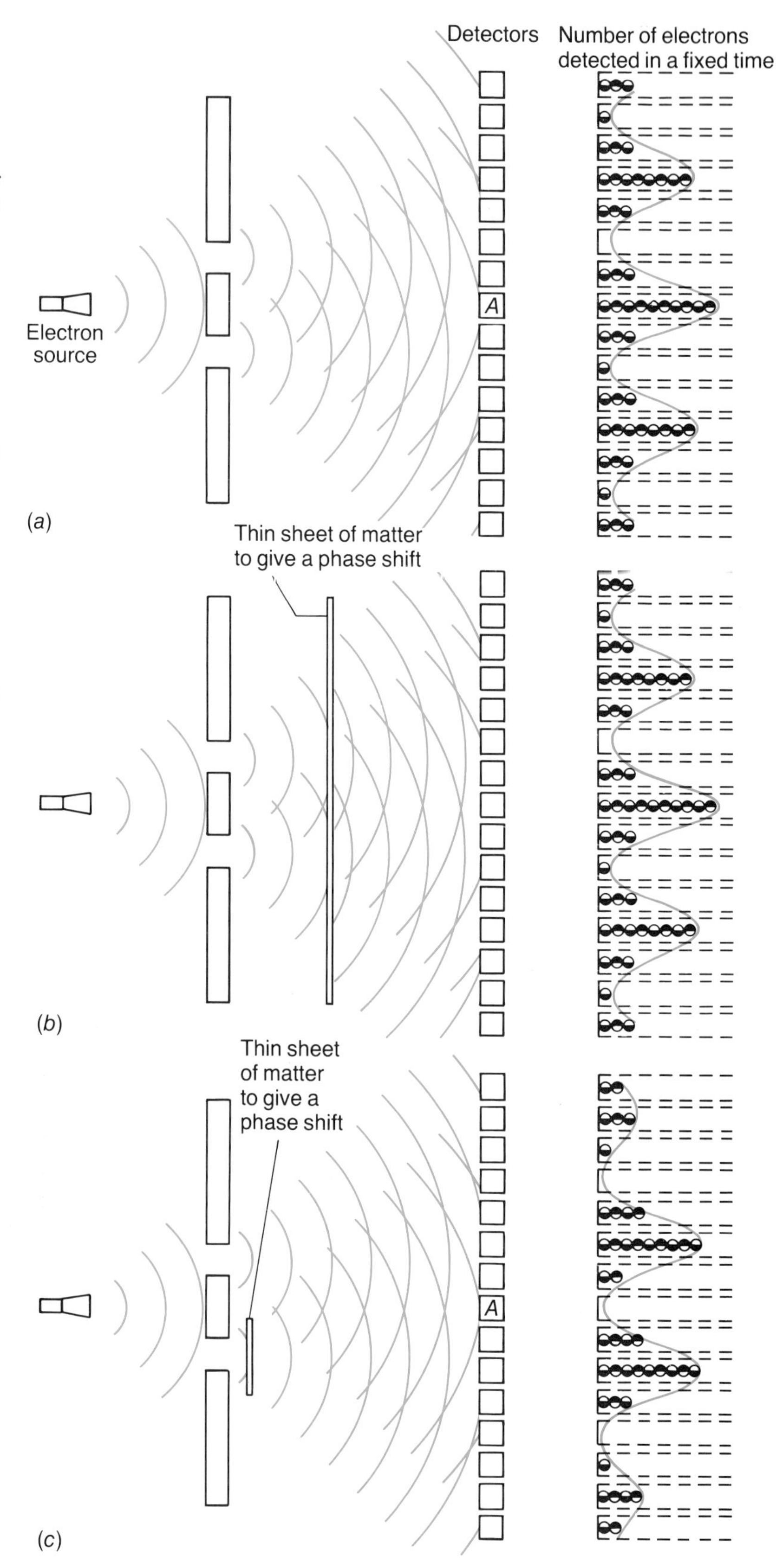

Fig. 10.1 The double-slit experiment with electrons – reprise: (a) A reminder of the interference pattern observed for the usual double-slit experiment with electrons. The electrons arrive individually at the detector and they are represented as black and white circles to remind us that we cannot tell which slit they passed through. Notice that the pattern is highest at the central detector marked A. (b) The interference pattern is unchanged if a thin sheet of matter is inserted between the slits and the detectors. The electron waves coming from both of the slits undergo the same change in phase. Thus, at the detector, the two waves will still add up to give a peak or still cancel to give a dip. We say that there is a 'global' phase invariance since the pattern is unchanged – 'invariant' – provided the sheet crosses the entire region behind the slits. (c) If a thin sheet of matter is inserted behind only one of the slits the pattern changes. Instead of the detector at A registering a maximum number of electron counts, it now records a dip in the interference pattern. The pattern has changed because the matter has altered the phase of only one of the electron waves. This shows that a 'local' change in the phase does not leave the interference pattern 'invariant'.

(d) The presence of a magnetic field will also cause the interference pattern to change. Apart from the fact that we cannot say which slit the electron passed through, this is more or less what we would expect, since in classical physics electrons are expected to be deflected by magnetic fields.
(e) The famous Bohm–Aharanov experiment showed that there was a shift in the interference pattern even if the magnetic field was shielded so that the electron paths from each slit to the detector did not pass through any magnetic field! The shielded magnetic field has been achieved in practice by using a long, thin electromagnetic coil, thinner in diameter than a human hair.
(f) The phase shift caused by the insertion of a sheet of matter behind one of the slits can be exactly compensated for by adjusting the magnetic field. This shows that 'local invariance' can be achieved provided the magnetic field interacts with the electron in just the right way. This is the basic principle behind all 'gauge' theories.

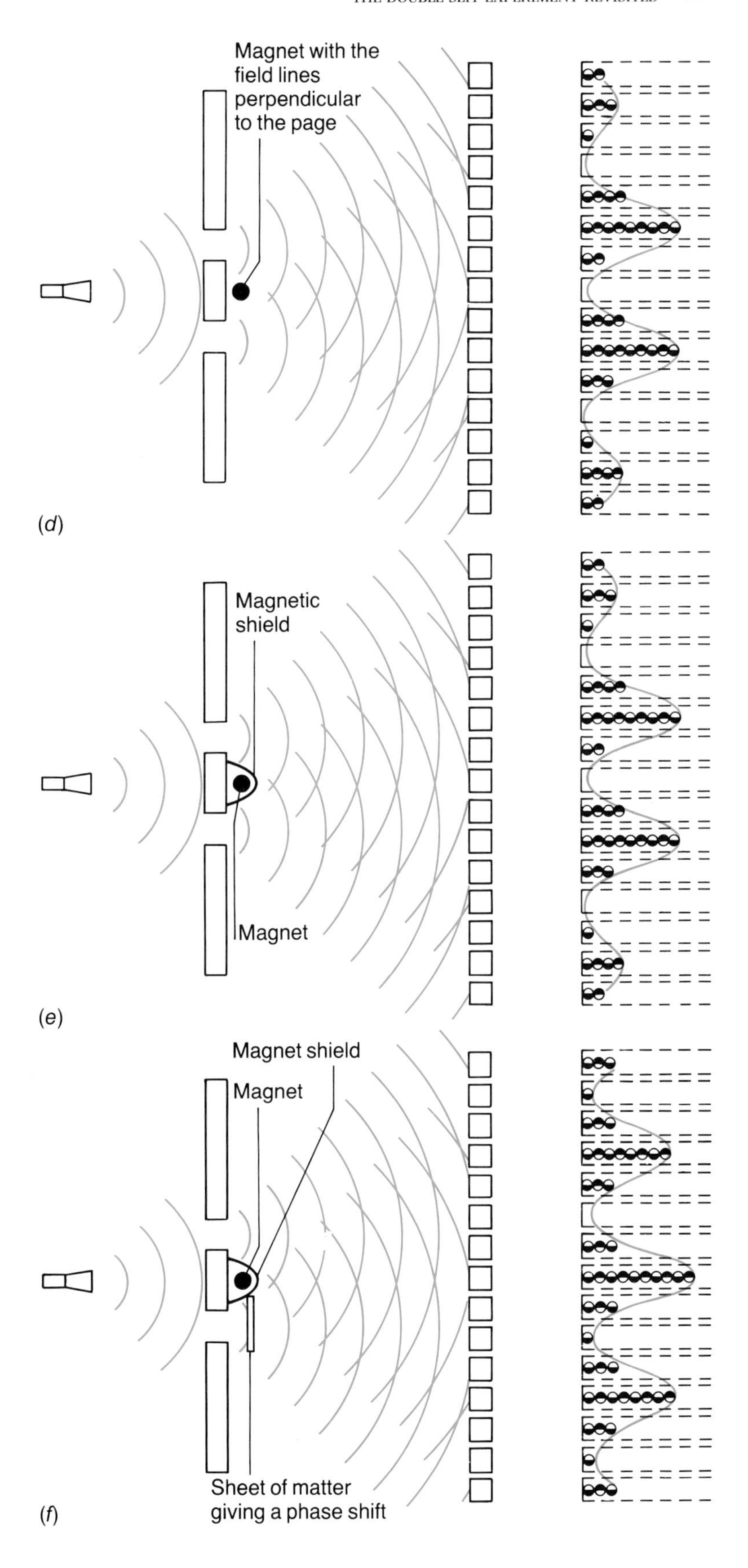

neutrons (see chapter 3), the matter in the sheet will interact with the electrons and alter the phase of the electron waves that pass through. The phases of the electron waves arriving at the screen will therefore be altered – where there was originally a crest there may now be a trough, and so on. However, the important point is this: since the phases of the waves from both slits are affected by the same amount, the interference pattern will not change. This is because at points on the screen where the two waves used to be in phase, the two waves will still be in phase and there will still be as many electrons arriving. Similarly, at points where they were originally out of phase, the two waves will still be out of phase. Thus, there will still be no electrons arriving at these points, exactly as in the experiment without the sheet of matter inserted. Since the interference pattern does not change when we insert the thin sheet of matter, physicists say that this is an 'invariance' of the double-slit experiment. More precisely, since the only relevant effect of the matter on the electrons is to cause a shift in phase in the electron waves, this is called a 'phase invariance'.

There is another feature of this invariance property that we want to stress. This type of 'phase invariance' requires that we insert a phase-changing sheet of material that spans the entire area of the screen. However, if we insert just a small piece of the material behind only one of the slits the interference pattern changes. This is because, in this case, the material will only affect the phase of waves from one of the two slits. Thus, at the screen, at points where we used to have two crests arriving, we may now have a trough arriving with a crest. We can summarize this as follows. If we only affect the phase of the electron waves in a small 'local' region, the interference pattern is changed by the insertion of the piece of material and there is no invariance. Only if we alter the phase everywhere, a 'global' phase change, does the interference pattern of the electrons remain unchanged so that we have an invariance. The double-slit electron interference experiment therefore has a 'global phase invariance'. It does not have a 'local phase invariance'.

Hermann Weyl (1885–1955) was an outstanding mathematician who also made important contributions to physics. At the height of his career in 1933, he resigned his post at the University of Gottingen in protest at the dismissal of his Jewish colleagues. Like so many other German scientists, Weyl went to the USA and became a member of the Institute for Advanced Study in Princeton, New Jersey. In the 1920s, when trying unsuccessfully to unify gravity and electromagnetism, he introduced some of the ideas of modern gauge theories. The term 'gauge theory' is a relic of these attempts: it would be far more appropriate to use the term 'phase theories' in the modern context.

To illustrate more graphically the difference between 'global' and 'local' effects Feynman once gave this example. Suppose we were interested in the total number of cats in the world at any given time. If we look at the population of cats for a short enough time so that no cats are born or die during this time, then the total number of cats remains constant. We can say that the number of cats is 'conserved'. But we know more than this. We know from experience that cats are conserved in a 'local' way. For example, if five cats disappear in Pasadena and reappear at the same instant in Southampton, that would be an example of 'global' conservation of cats. But we know that cats don't work like that! The number of cats in every small local region is conserved and it is this local conservation of cats that leads to the global conservation of the total number of cats.

This story has a more serious point to make. Physicists are always intrigued whenever they find some sort of invariance principle and they immediately try to see if they can find a better one. In particular, in the case of our double-slit experiment with electrons, we have found a 'global phase invariance'. How can we do better? Well, the necessity to have to change the phase of the electron wave everywhere, at the same time, in order to have an invariance seems a somewhat irksome and unnatural restriction. Would it not be more natural to have a theory that allowed us to make a phase change in some small local region without having to worry about what was happening elsewhere? In other words, is there some way of fixing things up so that we can allow some sort of local alteration of the phase and yet still have an invariance? There is – and the resulting theory is QED!

To see how the connection with QED comes about, we must first tell you what happens if we perform the double-slit experiment in the presence of a

Chen Ning Yang won the Nobel Prize with T. D. Lee in 1957 for predicting the violation of left–right symmetry by weak interactions. Earlier, in 1954, together with Robert Mills, Yang had written down a generalization of the gauge theory of ordinary electromagnetism. This was also done independently by a PhD student named Robert Shaw in Cambridge, UK, at about the same time. These 'Yang–Mills' theories were the precursors of modern gauge theories.

magnetic field. In fig. 10.1*d* we show the experimental set-up, with a magnet in position behind the slits. Since, even in classical electromagnetism, magnetic fields cause the trajectory of a charged particle to curve, it should be no surprise that the interference pattern changes. However, in terms of the quantum mechanical wave picture of electrons this is not so obvious. Since the interference pattern does change, the effect of the magnetic field must be to change the phase of the electron wave. This is all roughly in accord with our 'classical' expectation, but quantum mechanics has another surprise in store for us. Remarkably, if we arrange to shield the magnetic field so that no magnetic field can penetrate the region where the two electron waves travel and overlap, the interference pattern still changes! This phenomenon goes by the name of the 'Bohm–Aharanov' effect. It aroused much controversy amongst physicists until the effect was conclusively confirmed by experiment as recently as 1960.

It is the experimental observation of the effect of magnetic fields on the phase of electron waves that leads to the possibility of achieving a 'local' phase invariance. At this point we must come clean and confess that the precise details of how this comes about are too complicated to explain here. However, we can give some idea of the mechanism of this local phase invariance by means of the following observation. Suppose we insert a thin sheet of material behind only one of the slits. As we have argued, the interference pattern will change. Now consider what happens if, at the same time as this sheet is put into position, a magnet is inserted behind the slits. It is clearly plausible that we can arrange for the magnetic field to compensate for the effect of the phase change caused by the small sheet of matter. We would then observe no change from the original interference pattern. Thus, we have made a local phase change in one of the electron waves yet maintained the invariance of the interference pattern by introducing a magnetic field at the same time. The full story of the connection between local phase invariance and magnetic fields is a bit more subtle than it appears from this argument. Nonetheless, the conclusion suggested by this example is correct. It is only because electrons interact in a very specific way with a magnetic field that a 'local phase invariance' is possible.

This is the vital clue from QED that enables us to construct theories of the weak and strong forces. QED is a theory that has the interactions of electrons and electromagnetic photons carefully arranged to maintain a local phase invariance of the electron wavefunction. Unfortunately, for obscure historical reasons, physicists usually use the term 'gauge invariance' to describe this state of affairs, instead of the much more informative 'local phase invariance'. Hence, QED is known as a 'gauge' theory rather than what it actually is, namely a 'phase' theory. How does all this help us to find a theory of the weak or the strong interactions? The trick is to reverse the argument. In other words, suppose we did not know how electrons and photons interacted. If we now demand that any theory with electrons must have a local phase invariance, we would be forced to introduce magnetic fields and invent QED! This is all there is to the 'gauge principle': demanding local phase invariance determines the interactions of the theory. We will now use this beautifully simple idea to construct the theories that are believed to describe the weak and strong forces.

The birth of particle physics

Before we can describe the application of the gauge principle to weak and strong interactions, we must first give you a rapid survey of some of the most significant discoveries in 'particle physics' and also introduce some terminology. In 1932, when Chadwick discovered the neutron, all was simple: there seemed to be only three elementary building blocks of matter:

Hideki Yukawa won the 1949 Nobel Prize for his prediction of mesons as carriers of the strong force. Yukawa was the first Japanese scientist to win the Nobel Prize.

the proton, the neutron and the electron. The proton and neutron are much more massive than the electron and are called 'baryons', from the Greek word *barys* (βαρυς) meaning 'heavy'. The electron, on the other hand, is now known to be one of a family of particles known as 'leptons', from the Greek word *leptos* (λεπτός) meaning 'light' (as opposed to heavy). We have already met another type of lepton. This is Pauli's neutrino, the mysterious particle involved in the radioactive decay of a neutron into a proton, which we discussed in chapter 7. This classification into baryons and leptons seems a bit elaborate just to describe these four particles. Its usefulness only becomes apparent when we appreciate that, over the past 40 years or so, hundreds more 'elementary' particles have been discovered. Fortunately, after decades of confusion, a large measure of order has now been restored to elementary particle physics. As we shall see, our new understanding is due to a recognition of the leading role played both by quarks and by local phase invariance.

In the preceding chapters, we encountered Feynman diagrams as a pictorial way of representing interactions of particles. For example, a diagram that occurs in the scattering of electrons by a proton is shown in fig. 10.2. In the diagram, a virtual photon is 'exchanged' between the electron and one of the quarks that make up the proton. The concept of a force arising from the exchange of virtual particles gives us an idea of the distance over which this force can act. From the uncertainty principle we argued that we could 'borrow' energy ΔE for a time $\Delta t \approx h/\Delta E$ without spoiling energy conservation. If we multiply this time, Δt, by the velocity, v, of the particle, we obtain an estimate of the typical distance such a particle can travel

$$R = v \times (h/\Delta E)$$

$$\text{range} = \text{velocity} \times \text{time}$$

It was this argument applied to the known range of nuclear forces that led the Japanese physicist, Hideki Yukawa, to predict the existence of particles with a mass in between that of the electron and the proton.

Yukawa predicted the mass of this particle to be about 200 or 300 times heavier than the electron: a proton, remember, is about 2000 times heavier than an electron. This prediction was made in 1935, and no such particle had ever been observed. Not surprisingly, then, when particles with about the right mass were found in cosmic ray experiments two years later, this

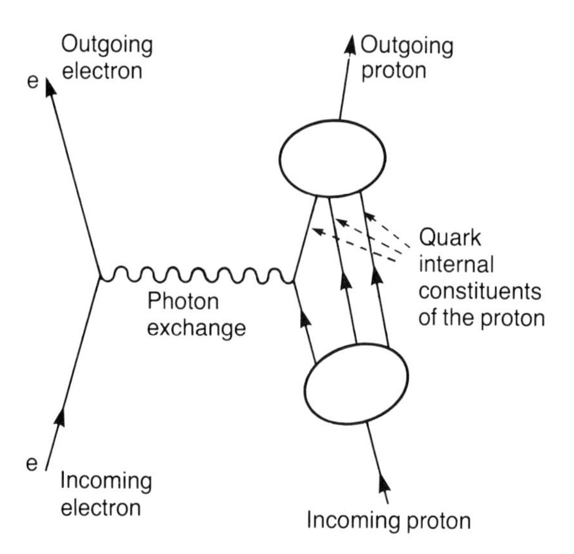

Fig. 10.2 Feynman diagram for electron–proton scattering. The scattering process is represented by the 'exchange' of a virtual photon between the electron and one of the quarks in the proton.

seemed to be dramatic confirmation of Yukawa's prediction. The war intervened to slow down research on these new particles but it did not stop research completely. Three young Italian physicists, Conversi, Pancini and Piccioni, were in hiding from the Germans to avoid being deported to forced labour in Germany. Working in a cellar in Rome, they discovered some very puzzling properties of these new particles. The particle did not behave at all like a 'carrier' of the strong force. Indeed, instead of interacting strongly with the nucleus, the particle seemed to interact more like an electron. The mystery was not cleared up until 1947 when it was suggested that there might be two new particles with about the same mass. One was the particle that had already been observed, which behaved like a 'heavy electron', while the other, yet to be observed particle was Yukawa's strong force carrier. This guess was shown to be correct when Powell and Occhialini in Bristol, UK, obtained cosmic ray tracks in photographic emulsion that conclusively established the existence of Yukawa's elusive particles. After some debate as to whether these new particles should be called 'yukons', in honour of Yukawa, these intermediate mass particles are now called 'mesons', from the Greek word *mesos* (μέσος) meaning 'in the middle'. The 'heavy electrons' discovered first are now called 'muons'. Yukawa's reasons for choosing to become a theoretical physicist rather than an experimental physicist provide an amusing footnote to all this. He has said that his decision was due, in part, to his 'inability to master the art of making simple glass laboratory equipment'!

The discovery of Yukawa's meson marks the birth of modern particle physics. It was the result of the development of new and better methods of observing the collisions of high energy particles. This search for new techniques continues to this day. Powell and Occhialini had been collaborating with the photographic laboratories of Ilford Ltd, to make better

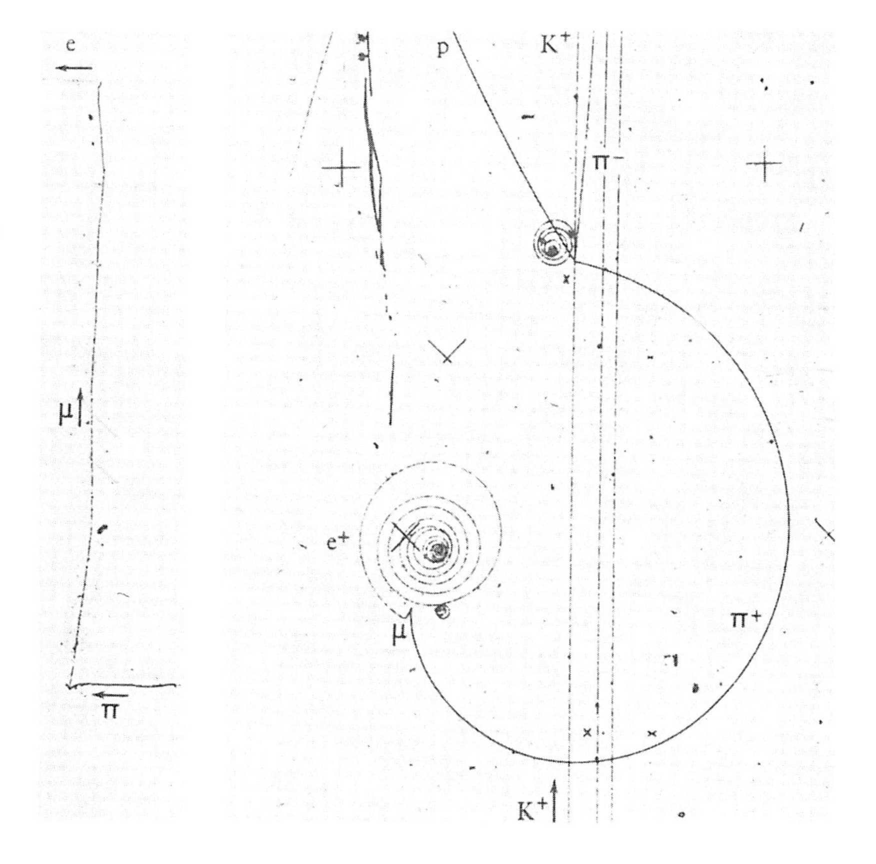

Fig. 10.3 Pions decay to a muon and an unseen neutrino. The muon in turn disintegrates into an electron and two more neutrinos. The left hand picture shows the tracks left by this decay chain in a photographic emulsion. On the right the same process is shown in a bubble chamber. Because of the magnetic field present in the chamber, the tracks are curved and the slow moving electron winds up like a watch spring.

Fig. 10.4 One of the neutrino experiments at CERN laboratory in Geneva. The detector weighs 1400 tons and consists of thick iron plates interleaved with scintillation counters and drift chambers to detect the charged particles coming from the neutrino interactions.

emulsions to show particle tracks. Occhialini then took some of these new plates and exposed them to high energy cosmic rays on top of a mountain in the French Pyrenees. What then happened to these plates is best described in Powell's own words.

'When they were recovered and developed in Bristol it was immediately apparent that a whole new world had been revealed. The track of a slow proton was so packed with developed grains that it appeared almost like a solid rod of silver, and the tiny volume of emulsion appeared under the microscope to be crowded with disintegrations produced by fast cosmic ray particles with much greater energies than any which could be generated artificially at the time. It was as if, suddenly, an entry had been gained into a walled orchard, where protected trees had flourished and all kinds of exotic fruits had ripened undisturbed in great profusion.'

Nonetheless, even with the great theoretical advances made in particle physics over the last decade, the reason for the existence of the muon remains a mystery. The Nobel Prize winner, Isidor Rabi, is reported to have said 'Who ordered that?' when told of the muon's discovery, and his question is still unanswered. We have, however, uncovered a further clue which may yet prove vital for solving the puzzle of the leptons. This discovery also involved a curious repetition of the confusion surrounding the discovery of the muon and Yukawa's meson. In the mid 1970s, physicists were looking for a new meson to confirm their theories about a new type of quark (see below). Instead, at about the same mass expected for this new meson, another heavy electron was identified. This discovery was largely due to the efforts of the US physicist, Martin Perl, who named the new lepton the 'tau'. The expected new mesons were found shortly afterwards, thus completing a very strange and still unexplained re-run of history.

Yukawa's meson was called a 'pi-meson' or 'pion' for short. Other strange cosmic ray events were discovered shortly afterwards (see fig. 10.5). The characteristic feature of these peculiar new events was the existence of two 'vees' pointing back to the initial interaction point. Since only charged particles leave tracks in the detector, we deduce that the two vees are the charged decay fragments of two neutral particles created at the point of the original collision. These neutral particles, soon to become known as 'strange' particles, then travel some distance before decaying. By taking photographs of these events, in the presence of a magnetic field, and making careful measurements of the curvature of these tracks, we can apply the laws of energy and momentum conservation to deduce the masses of all the

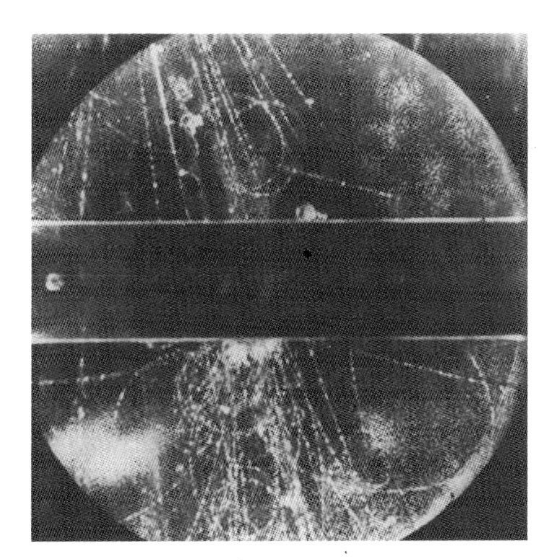

Fig. 10.5 The discovery of strange particles. A neutral K meson is produced by a cosmic ray interaction in the lead plate that passes through the cloud chamber. Its decay products are charged pions and can be seen as a 'vee' in the lower right of the picture.

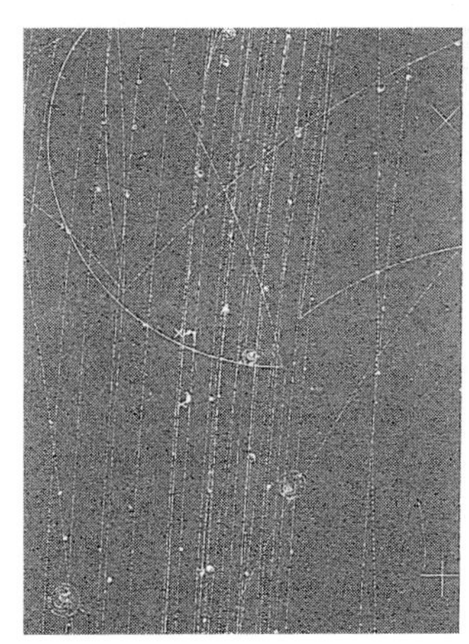

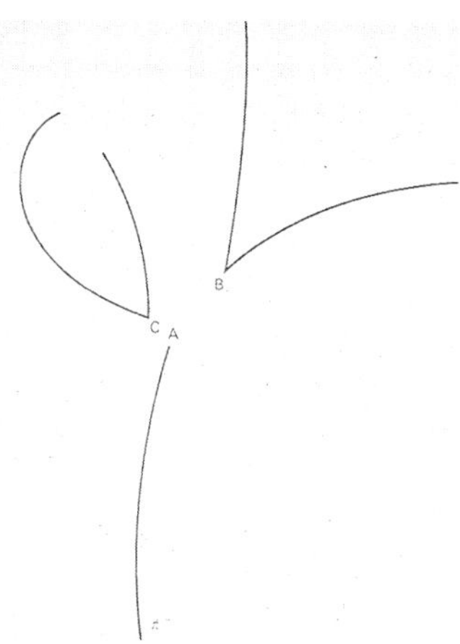

Fig. 10.6 Double 'vee' events are common in bubble chambers. A negatively charged pion collides at A with a proton of the hydrogen that fills the chamber. In the reaction two strange particles are produced, a neutral K meson and a neutral Λ baryon. The Λ decays at B to a proton and a π^-, and the kaon decays at C to a π^+ and a π^-.

particles participating in the event. In this way, new 'strange' baryons and mesons were identified. Why were these new particles called strange particles, apart from the obvious strangeness of their production in pairs of vees? Consider a typical 'double vee' event (see fig. 10.6). This corresponds to the reaction

$$\pi^- + p \rightarrow \Lambda^o + K^o$$

where the lambda (Λ) is a strange baryon and the kaon (K) a strange meson. The most puzzling thing about these strange particle events was that, while it was easy to create pairs of strange particles from collisions of pions and protons, these strange particles, left to themselves, showed a marked reluctance to turn back into protons and pions. In other words, the production of pairs of strange particles takes place via the strong interactions but the decay of individual strange particles is governed by weak interactions

$$\Lambda^o \rightarrow p + \pi^-$$

$$K^o \rightarrow \pi^+ + \pi^-$$

This has been confirmed by observation of other obviously weak decay 'modes' such as

$$\Lambda^o \rightarrow p + e^- + \bar{\nu}$$

$$K^o \rightarrow \pi^- + e^+ + \nu$$

We now know that strange particles possess a new type of 'charge' that distinguishes them from 'ordinary' matter such as protons, neutrons and pions. In the strong interaction production reactions, the initial and final states must always have the same strangeness. Thus, in our example above, since the kaon has strangeness $+1$ and the lambda particle strangeness -1, the final state has total strangeness zero as does the pion–proton initial state. In the decays of strange particles, however, the strangeness does not balance on each side of the reaction. This means that the process is not allowed to proceed via the fast strong interactions, but can only take place, reluctantly, by the much slower weak interactions of beta radioactivity.

Fig. 10.8 Experimental particle physics now involves collaborations of large numbers of physicists from many countries. Collected round the bubble chamber in this photograph are some of the 114 people who shared in the hunt for the Ω^-.

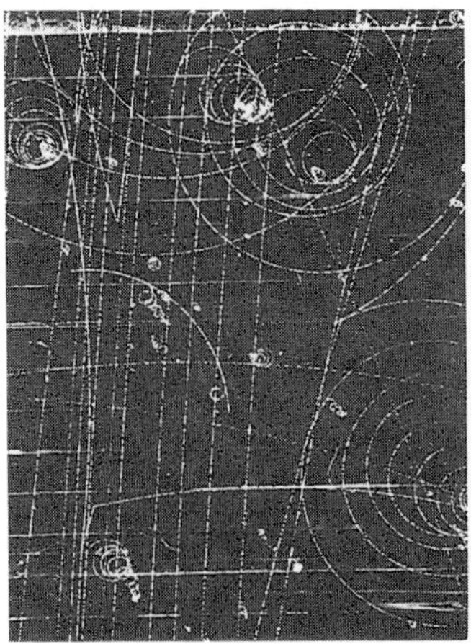

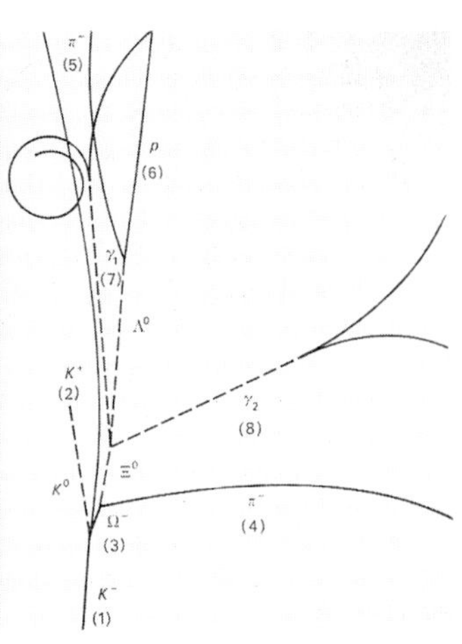

Fig. 10.7 The world's first Ω^- event. The Ω^- contains three strange quarks and decays via a three stage process to a Ξ^0 and then to Λ^0 before finally shedding the last of its strangeness in the Λ decay. A neutral pion is also created in the decay chain which itself decays to two photons. The photons are neutral and leave no track so reconstruction of what happened would normally be very difficult. This event is remarkable because both of the photons from the pion decay have been converted to electron–positron pairs in the bubble chamber. This was the happy chance that enabled Brookhaven to win the race to find the Ω^- particle. Lightning seems to have struck twice in the same place, for Brookhaven were similarly fortunate in their finding of a charmed baryon many years later!

During the 1950s and 60s experimentalists also found many short-lived 'excited states' of the proton, neutron and pion, as well as of the strange particles. How could all these particles be fundamental? Gell-Mann and Zweig brought some sense to all this with the introduction of quarks. Baryons are made up of three quarks, and mesons of a quark and an antiquark. Dramatic confirmation that Gell-Mann was on the right track had come a year or so earlier with his prediction of a particle called the Ω^-. The proton and neutron are made up out of the two types of 'non-strange' quarks: the Ω^- is composed of three 'strange' quarks. The short-lived excited states could now be understood as excited states of these quark systems, much in the way that we understand excited states of atoms and nuclei. At the beginning of the 1960s it was fashionable to suppose that all particles were equally elementary and 'nuclear democracy' was the watchword. It took physicists some time to come round to the idea that there really were fundamental constituents of matter. During the late 1960s a few perceptive physicists, like Dick Dalitz in Oxford, UK, persevered with the idea of explaining the excited states in terms of quarks, often in the face of ridicule and disbelief from some of their colleagues. But at the end of the 1960s, the quark picture of elementary particles received dramatic confirmation from the electron–proton scattering experiments performed at Stanford, California (see chapter 3). These experiments had a natural explanation in terms of the electron scattering off quarks inside the proton, and later neutrino experiments at CERN have confirmed this picture. It is now generally accepted that both baryons and mesons contain quarks – despite the fact that no-one has ever seen a free, isolated quark.

We conclude this section by introducing you to yet another new word. Both baryons and mesons interact via the strong nuclear force. Leptons, on

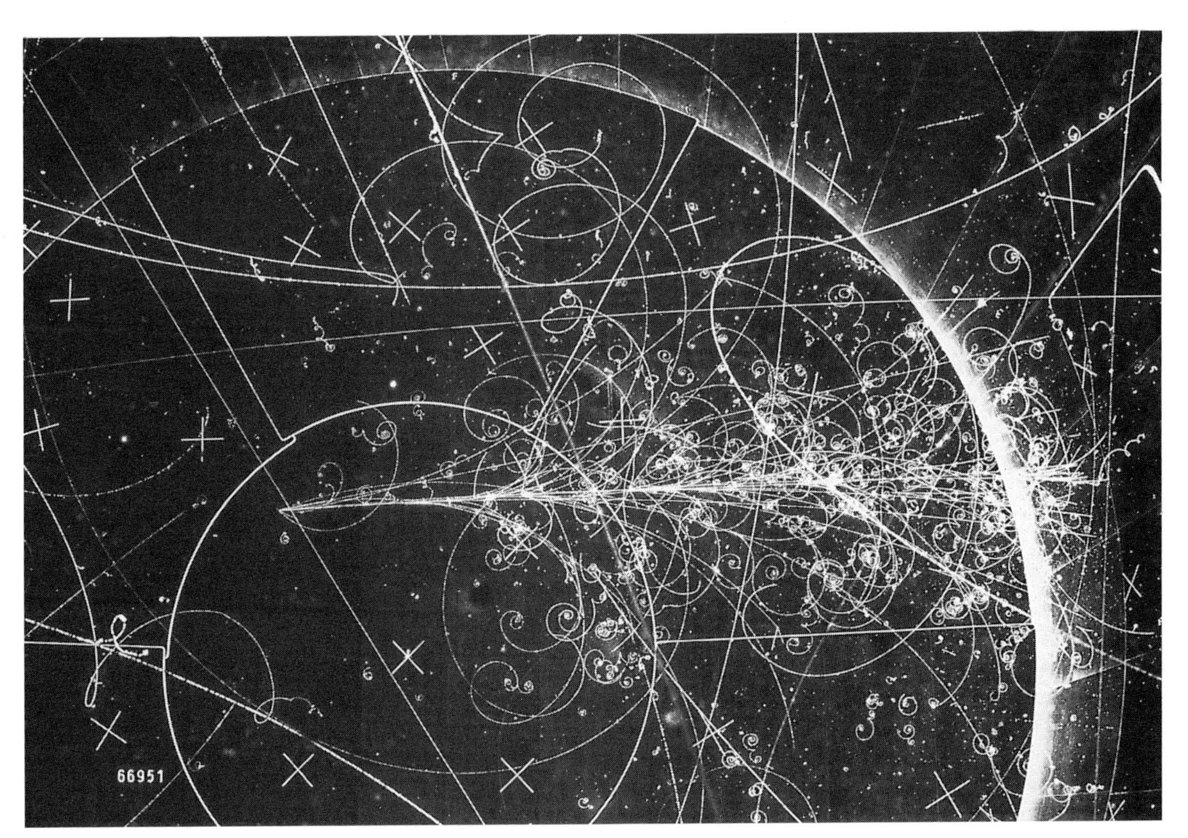

Fig. 10.9 A photograph of a neutrino reaction taken in BEBC – the Big European Bubble Chamber at CERN. The neutrino beam enters from the bottom and collides with a quark in the proton to produce a very complicated spray of particles.

the other hand, 'feel' only the weak and electromagnetic forces. Particles that interact via the strong force are called 'hadrons'. This word was first coined by a Russian physicist, Okun, from a Greek word that at first sight does not seem very appropriate, since *hadros* is the Greek word for 'bulky'. However, *leptos*, from which lepton is derived, has another meaning besides 'light', namely 'fine-grained'. It is in this sense that *hadros* (ἁδρός) is the opposite of *leptos* and therefore quite appropriate to distinguish between particles that feel or do not feel the strong nuclear force.

Sheldon Glashow (left) and Steven Weinberg shown together at a press conference in Harvard on the day they won the Nobel Prize. Together with Abdus Salam, they were awarded the prize for contributing to the unification of the weak and electromagnetic forces in a single theory.

Weak photons and the Higgs vacuum

The 1979 Nobel Prize for physics was awarded to three physicists, Sheldon Glashow, Abdus Salam and Steven Weinberg, for their 'contributions to theory of the unified weak interaction between elementary particles'. This was a bold move by the Nobel Prize committee since the unified theory of Glashow, Salam and Weinberg predicted the existence of two new particles, the W and Z, with masses 80 or 90 times heavier than the proton. At the time, no such particles had been observed. The committee must have heaved a sigh of relief when these predictions were spectacularly verified at the proton–antiproton collider at CERN in Geneva. Carlo Rubbia and Simon van de Meer were awarded the 1984 Nobel Prize for their part in making these experiments possible. How does the 'unification' of weak and electromagnetic interactions come about? And what has all this got to do with the gauge principle? To understand the answers to these questions we must look again at Yukawa's argument about the range of forces and see how this is related to the mass of the virtual particle being exchanged.

Yukawa was able to deduce the mass of the pion from the observed range of nuclear forces. The heavier a particle is, the more energy that has to be

Abdus Salam was born in what is now Pakistan, and he studied mathematics at Lahore University. Originally intending to become a civil servant, he eventually ended up on a scholarship to Cambridge, UK, studying physics. Salam is now one of the most prominent scientists of Islamic faith. He donated his share of the Nobel Prize to his institute in Trieste, Italy, which encourages scientists from the developing countries.

borrowed to create it, and the shorter the distance it can travel on 'borrowed time'. Now the energy E, momentum p, and mass m, of a particle moving at very high, 'relativistic' speeds are related by the equation

$$E^2 = p^2c^2 + m^2c^4$$

where c is the velocity of light. At low, non-relativistic velocities this equation reduces to a more familiar result

$$E = (p^2/2m) + mc^2$$

This says that the total energy of a non-relativistic particle is just the usual kinetic energy plus Einstein's mass energy. For a photon, however, we must use the relativistic formula since photons always travel with the velocity of light. Moreover, photons are found to have 'zero mass' which just means that the energy and momentum of a photon are related according to the first equation above, with m set to zero. If we now go back to Yukawa's 'borrowed energy' argument, this means that we can create virtual photons with very low momentum and almost zero total energy. Such virtual photons can travel almost as far as they like without getting into trouble with the energy–time uncertainty relation. We therefore expect electromagnetic interactions to be effective over very large distances, and this expectation is confirmed by experiment.

At first sight, the requirement of local phase invariance seems to require that 'gauge particles' must have zero mass, like the photon. This is because we need to be able to fix up the effects of a local phase change over all positions on the screen and this can involve very large distances. In fact, this turns out not to be the case, but massive gauge particles are only possible in a rather peculiar way. We can illustrate this by looking again at magnetic fields and superconductors. In chapter 8 we saw that magnetic fields do not penetrate far inside a superconductor. On entering the superconductor the magnetic field falls off very rapidly over a very short distance. The effect is caused by currents that are set up within the superconductor when it is placed in a magnetic field. These currents produce magnetic fields that tend to 'screen out' or cancel the applied magnetic field inside the metal. This 'diamagnetic' effect takes place in all metals, but in a superconductor, because there is no electrical resistance, these induced currents produce a magnetic field that completely cancels out the applied magnetic field inside the metal, apart from in a very thin surface layer. Consider this situation in terms of the range of the magnetic field inside the superconductor. Since the field only penetrates a very short distance, it is as if, inside the superconductor, the photon has acquired a very large mass.

In this case, of course, we known that this effective photon mass is caused by the superconducting 'screening' currents induced by the applied magnetic field, and that outside the metal photons are massless. But now try and imagine what the world would look like from the point of view of someone small enough to live permanently inside such a superconductor. Such 'people' may not be clever enough to realize that they are living in the presence of 'vacuum' screening currents. Instead, they would deduce that photons have a mass related to the distance magnetic fields can travel in the metal. It is only in this sense that 'gauge particles' can have a mass and still preserve the local phase invariance.

What has this got to do with weak interactions? In the previous chapter we drew Feynman diagrams for electron–quark scattering with the electromagnetic interaction mediated by the exchange of a virtual photon. For weak interactions we can draw similar diagrams. In the beta decay of the neutron, for example, a 'down' quark changes into an 'up' quark emitting a virtual W particle which decays into an electron and an antineutrino. However, unlike the electromagnetic interaction, the range of the weak

Fig. 10.10 The Big European Bubble Chamber (BEBC) at CERN. This can be filled with liquid hydrogen or a mixture of neon and hydrogen. The chamber is surrounded by superconducting niobium–titanium coils which produce a very large magnetic field inside the chamber.

Fig. 10.11 The telex sent by the Nobel Prize Committee to Abdus Salam informing him that he had won the Nobel Prize.

T
261503 IMPCOL G
22931 GNTC G
D970 LA989 UOP411
GXXX CO SWSM 094
STOCKHOLM 94/89 15 1145 PAGE 1/50

PROFESSOR ABDUS SALAM
IMPERIAL COLLEGE OF SCIENCE
AND TECHNOLOGY
PRINCE CONSORT ROAD
LONDON(SW7 2AZ)

DEAR PROFESSOR SALAM,
I HAVE THE PLEASURE TO INFORM YOU THAT THE ROYAL SWEDISH ACADEMY
OF SCIENCES TODAY HAS DECIDED TO AWARD THE 1979 NOBEL PRIZE
IN PHYSICS TO BE SHARED EQUALLY BETWEEN YOU, PROFESSOR SHELDON L.
GLASHOW AND PROFESSOR STEVEN WEINBERG, BOTH AT HARVARD UNIVERSITY,
FOR YOUR CONTRIBUTIONS TO THE THEORY OF THE UNIFIED WEAK AND
ELECTROMAGNETIC INTERACTION BETWEEN ELEMENTARY PARTICLES,
INCLUDING INTER ALIA THE PREDICTION OF THE WEAK NEUTRAL CURRENT.
C.G. BERNHARD
SECRETARY GENERAL

IV SENT 1200 JC
22931 GNTC G
261503 IMPCOL G

T

Fig. 10.12 A view of the enormous UA2 detector installed in Underground Area 2 at CERN. The apparatus is placed in the SPS tunnel and protons and antiprotons circulating round the SPS are brought together to collide inside the detector.

Fig. 10.13 A mechanical mole bored a 7 km tunnel underneath the French–Swiss border near Geneva. The ring of magnets that forms the accelerator called the super proton synchroton (SPS) is located in this tunnel.

force is very small. By Yukawa's argument, therefore, the mass of the W must be rather large. The W particle must also be charged – unlike the photon which is electrically neutral. In fact, there appears to be little similarity between the local phase theory of QED, with its massless neutral photon, and any theory of weak interactions, which must involve massive charged W's.

It is here that the relevance of our discussion of the superconductor becomes apparent. Imagine that the 'vacuum' in which we live is analogous to a 'weak superconductor'. If this is the case then 'vacuum' screening currents can make it appear that the W particles have mass. This is the key idea behind the 'Higgs vacuum'. Perhaps not surprisingly, given the close relation to superconductivity, this mechanism for giving mass to gauge particles was first suggested by Philip Anderson, the distinguished 'solid-state' physicist whom we encountered briefly in chapter 8. In a superconductor, the screening currents are due to circulating 'Cooper pairs' of electrons. In the case of a gauge theory of weak interactions, these currents are believed to be due to particles which have become known as Higgs bosons. Peter Higgs is a British theoretical physicist, based in Edinburgh, who was one of the first to work out Anderson's ideas in a relativistic context.

Fig. 10.14 The European Nuclear Research Centre (CERN) located outside Geneva, Switzerland. The large circle marks the position of the underground SPS accelerator. Lake Geneva is visible towards the top left of the photograph and Mont Blanc is just visible on the horizon.

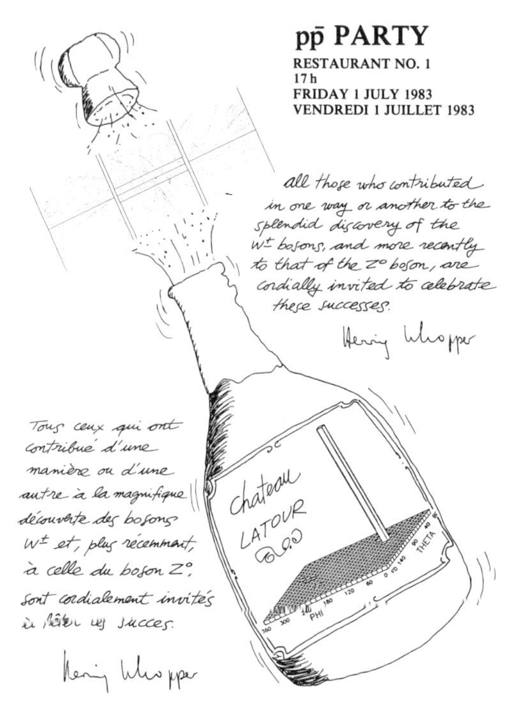

Fig. 10.15 A poster for the party held at CERN to celebrate the discovery of the W and Z bosons.

Why did the Nobel Prize committee have enough confidence to award the prize to Glashow, Salam and Weinberg before the W was discovered? One of the reasons was the successful prediction of a new type of quark – the so-called 'charmed' quark. This came about as follows. The theory that is now known as the 'standard model' or the 'GSW' model of electroweak interactions predicts that, in addition to the charged W particles, there should exist a neutral heavy Z particle – a genuine 'weak photon'. If this particle exists, then it must contribute to neutrino scattering via Feynman diagrams like that shown in fig. 10.17. Unlike diagrams involving W exchange, the quark charge is unchanged in such Z exchange diagrams. These reactions correspond to the 'neutral currents' referred to in the telex from the Nobel committee to Salam. After many rumours and false alarms, the discovery of such events was finally announced to the world at an international conference in London, UK, in 1974. The invited review speaker on weak interactions at the conference was a Greek physicist named John Iliopoulos. In his talk he threw down a famous challenge. Just as the discovery of neutral currents was the sensation at the London conference, so Iliopoulos offered to bet anybody a case of wine that the sensation of the next conference would be the discovery of the 'charmed' quark. Iliopoulos won his bet.

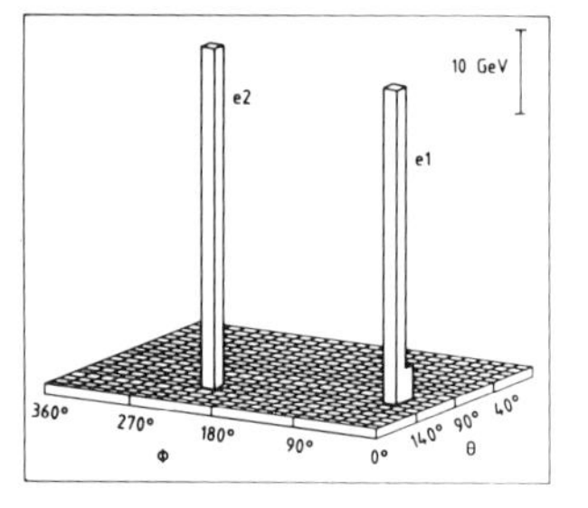

Fig. 10.16 A dramatic view of a Z event. The height of the column indicates the energies of the electron and positron from the Z decay.

Exactly how the existence of neutral currents implies the existence of a new type of quark is quite a complicated story. Nevertheless, if one believes in a gauge theory of weak interactions, a fourth quark is necessary in order to avoid a conflict with well-established experimental data. This new quark must have a new quantum number which Glashow called 'charm'. Just as the electromagnetic force is different for particles with different electrical charges, so the strength of the weak force depends on both the strangeness and charm of the quarks. This was the situation in the summer of 1974, and it is probably fair to say that not many physicists thought it likely that Iliopoulos would win his bet. There was great consternation in the physics community, when, in the autumn of that year, a spectacular new meson was discovered. Nowadays, nobody seriously doubts that this meson is made up from a charmed quark bound to a charmed antiquark. New mesons

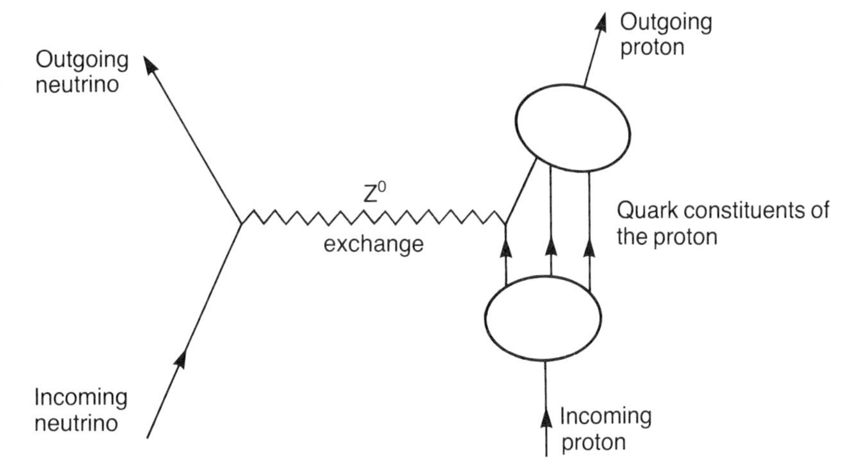

Fig. 10.17 Feynman diagram for neutrino–proton scattering. The neutrino exchanges a virtual Z boson with one of the quarks in the proton.

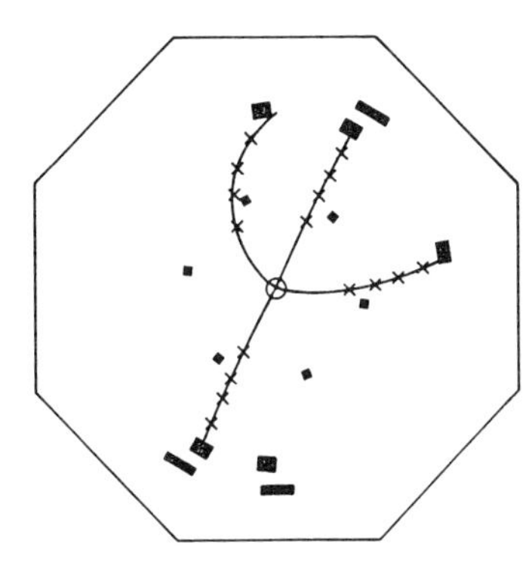

Fig. 10.18 The J/ψ particle was discovered more or less simultaneously at Brookhaven, New York, and at SLAC in California. Here is a reconstruction of a 'ψ looking' event from the electron–positron collider at SLAC. It is caused by a heavier version of the ψ decaying to the usual ψ and positively and negatively charged pions. The ψ is identified by its electron–positron pair decay products.

containing charmed quarks bound to uncharmed antiquarks have also now been found.

A final postscript to this success story is in order. At the time that Glashow, Salam and Weinberg made their contributions to the standard model of the electroweak interactions, there was a serious problem. Although their theory looked like a promising candidate to describe the experimental data, nobody knew how to do calculations beyond the 'tree' diagrams – diagrams with no closed loops. 'Loop' graphs usually involve more powers of the charge, e, that governs the strength of the coupling of the W's and Z's to quarks and leptons. Since e^2 is found, from experiment, to be very small, e^4 will be much smaller, and these 'higher order' diagrams should be relatively unimportant. Unfortunately, all attempts at calculating the effects of these loop diagrams had ended in failure and no-one knew what to make of these theories. It was not until a young Dutchman named Gerard 't Hooft came along that all became clear. In the words of the physicist Sidney Coleman, ''t Hooft's work changed the Weinberg–Salam frog into an enchanted prince'. Some years earlier, Coleman had reproached Tini Veltman, 't Hooft's thesis advisor, for persisting in his research in 'sweeping out a forgotten corner of theoretical physics'. It is fortunate that Veltman held firm against the prevailing fashions of the time and was one of the first to recognize the importance of gauge theories.

What remains to be done? Despite the succession of experimental

Samuel Ting with other members of his group that discovered the J/ψ particle in Brookhaven, New York. This particle is now thought to be a bound state of a charmed quark and its antiparticle. Ting shared the 1976 Nobel Prize with Burton Richter, who led the team that discovered the same particle in SLAC, California.

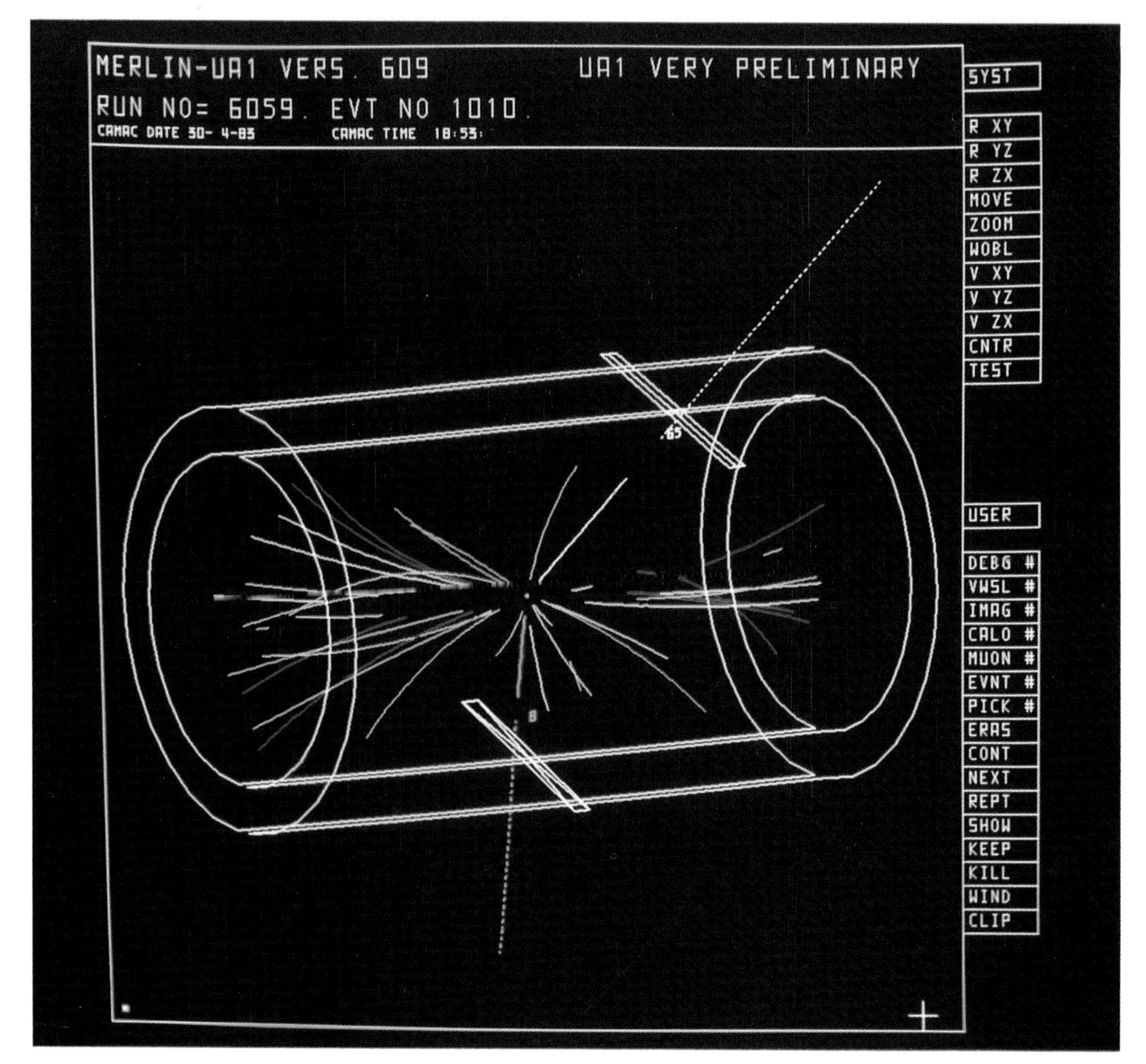

Fig. 10.19 A spectacular colour graphics reconstruction of the production of a Z weak boson in a proton–antiproton collision. The Z is identified by the blue and white tracks emerging from the sides of the cylindrical detector. These are left by the electron and positron coming from its decay. This event was seen in the UA1 experiment led by Carlo Rubbia.

triumphs for the standard model, there are still many unanswered questions. The model can accommodate the muon and the tau leptons but has no convincing reason for their existence nor any prediction for their masses. Nor do we have a real understanding of the masses of the various quarks, or why there now appear to be six varieties of quarks. The large electron–positron accelerator, 'LEP', at CERN will be able to study the Z particle in great detail and may provide answers to some of these questions and pose some new ones. There is, however, one crucial piece of the jigsaw that probably cannot be tested at LEP. This concerns the existence or otherwise of the mysterious Higgs particle. Any experimental search for this particle is hampered by the fact that the model makes no prediction for the mass of the Higgs particle. Moreover, the solid-state analogue of the Higgs particle in a real superconductor is a Cooper pair of electrons. Thus, the Higgs particle may not be a genuine elementary particle but may turn out to be composite. In any event, it is believed that something must show up at energies of around a million million electron volts (TeV). For this reason, US physicists now have ambitious plans to build a machine capable of attaining such energies.

A photograph of the leaders of the collaboration of physicists that discovered the J/ψ particle in November, 1974, at the electron–positron collider at the SLAC, California. From left to right they are Gerson Goldhaber, Martin Perl and Burton Richter.

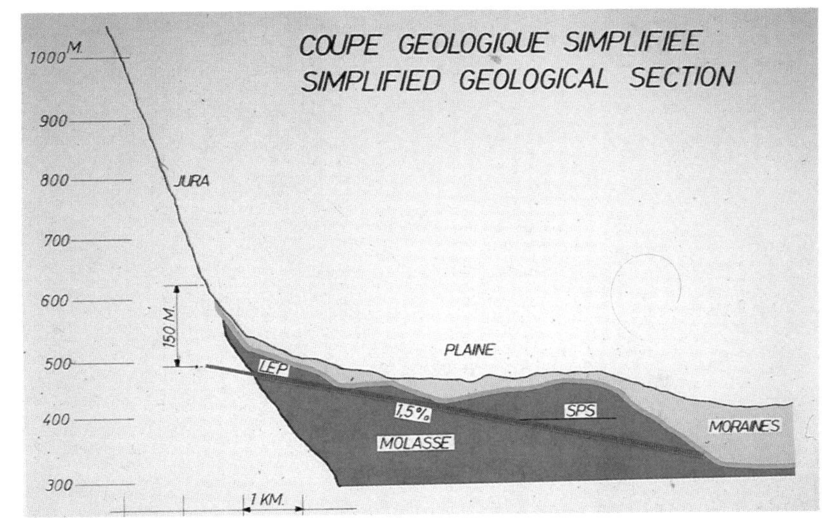

Fig. 10.20 A cross-section through the mountains near Geneva, showing the location of the large electron positron (LEP) accelerator ring currently under construction.

Quarks and gluons

In the early days of nuclear physics, physicists hoped that the theory of the strong force would be simple and elegant. With the discovery of the pion and all the other hadrons, together with all their excited states, it rapidly became apparent that the force between neutrons and protons was very complicated. However, during the time that physicists were discovering all these new particles, they also learnt that hadrons are built out of quarks. If there is to be a simple theory of hadronic forces, it is natural to look for an explanation in terms of quarks. Perhaps the so-called 'strong interactions' are merely a feeble shadow of enormously powerful inter-quark forces described by a simple and elegant law?

We have seen that quarks come in several different varieties: non-strange, strange, charmed and so on. However, it is the electroweak force that distinguishes between these different 'flavours' of quarks; the strong force is the same whether it acts on a strange or a charmed quark. At this point we must apologize for the somewhat 'jokey' names that particle physicists give

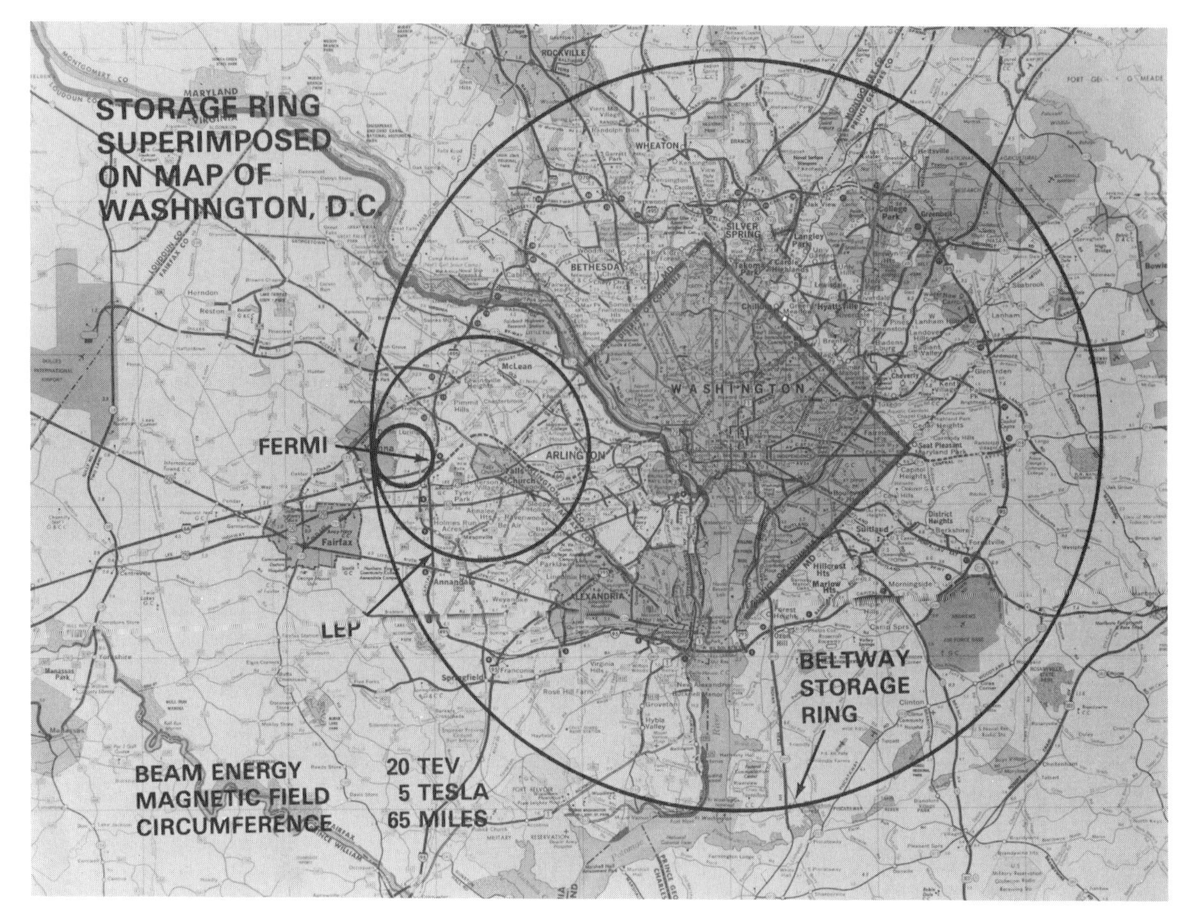

Fig. 10.21 The proposed proton–antiproton collider in the USA could completely encircle Washington, DC. The sizes of the LEP accelerator and the Fermilab accelerator are shown for comparison.

to these new quantum numbers. A quantum number like strangeness is a well-defined physical quantity. In fact, early on, some physicists used to refer to an equivalent quantity called 'hypercharge' instead of strangeness. Hypercharge certainly invokes a more formal and imposing image for particle physics, but most physicists prefer to use strangeness. Similarly, the non-strange quarks, as the name implies, have no strangeness but have different electric charges. Instead of referring to these quarks by their 'eigenvalues of the third component of isotopic spin', physicists prefer the shorthand 'up' and 'down'. Given that the first three quarks are called 'up', 'down' and 'strange', it should not come as much of a surprise that the next three quarks are called 'charm', 'top' and 'bottom'. This is not all an elaborate joke at the tax-payers' expense – it merely shows that physicists are human!

The strong forces do not notice the different 'flavours' of quarks. Instead, they are sensitive to yet another type of 'charge' carried by all the quarks. Inevitably, particle physicists refer to this new quantum number as 'colour', but again it should be remembered that this is only a shorthand for a very specific mathematical property. We could say, with pedantic accuracy, that 'quarks transform according to the fundamental representation of the special unitary group $SU(3)$'; it is surely preferable to say they carry a 'colour charge'. We can make the necessity for this colour quantum number somewhat more physical by the following argument. Consider the particle called the Ω^- that was predicted by Gell-Mann. This is a baryon and contains three quarks. We have focussed on this particle because its composition is particularly simple and can serve to highlight the problems

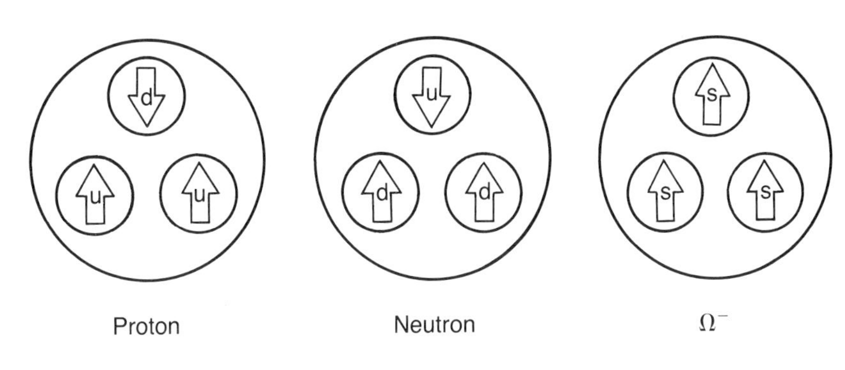

Fig. 10.22 The quark content of the proton, neutron and Ω^-. For the proton and neutron we have indicated possible quark spin orientations adding up to a net spin of 1/2. The Ω^- has spin 3/2 so that all three strange quarks must have their spins pointing in the same direction. This would be forbidden by the Pauli exclusion principle unless the quarks had some extra hidden quantum number.

we would have without a colour quantum number. In order to make up its electric charge of -1 unit and have strangeness -3, the three quarks must all be 'strange' quarks. The spin angular momentum of the Ω^- is also special in that it is 3/2. Since the Ω^- is one of the lowest-lying baryons the quarks are all in the lowest energy level with orbital angular momentum zero. Thus, the spin of the Ω^- must all come from the spin of the quarks. Roughly speaking, each of the quarks has spin 1/2 and all of these spins must point in the same direction to add up to a total of 3/2. This all looks very satisfactory – where is the problem? The problem lies with the Pauli exclusion principle that we discussed in chapter 6. The quarks are 'fermions' and must obey the Pauli principle. As things stand, all of the quarks in the Ω^- have the same quantum numbers and Pauli does not allow this. The colour quantum number of the quarks solves this problem. Colour has something to do with a mathematical construct known as a 'group', and, in particular, the group '$SU(3)$'. The 'threeness' of this group means that there are three different possible states for the quark. Again we usually refer to this situation more informally by talking about quarks coming in three different 'colours'. It is important to remember that this is just a shorthand for the mathematics: quarks do not have real physical colours that control the strong forces. Having said all this, we can now see how colour solves our problem with the Ω^-. There are three different possible colours for the quarks, so each quark in the must have a different colour – 'red', 'green' and 'blue', say – to satisfy the exclusion principle.

We are now in a position to describe the ingredients that make up the theory called 'quantum chromodynamics', the long sought theory of the strong force. This is a gauge theory based on the local phase invariance of the colour properties of the quantum amplitudes of the quarks. Although this may sound intimidating, it is difficult to imagine that the theory of the strong interactions could be simpler. Just as the electromagnetic forces are 'mediated' by zero mass gauge particles – the photons we have met so often – so we expect that the quark–quark interactions are described in terms of the exchange of similar 'strong photons'. Needless to say, physicists have given these particles the name 'gluons', because, in a very real sense, they are the glue that holds everything together. Photons couple to the ordinary electric charge of the quarks: gluons couple to the 'colour charge' of the quarks. Moreover, the gluons themselves carry a colour charge and the gauge principle dictates that, unlike photons, gluons must interact with themselves. Physicists believe that it is this key feature that makes quantum chromodynamics (QCD), where 'chromo' is for colour, so different from quantum electrodynamics (QED). Why do we say that QCD is 'so different' from QED? This is because it is easy for us to observe electrons in the laboratory but no-one has ever been able to observe a quark all by itself. Quarks have only been 'seen' in combination with other quarks and antiquarks inside hadrons. Most physicists believe that this is not an

Gerard 't Hooft was born in 1947 and is now professor of physics at the University of Utrecht in the Netherlands. While studying for his PhD under Tini Veltman in Utrecht, 't Hooft made a vital breakthrough in discovering how to make consistent Feynman diagram calculations for gauge theories.

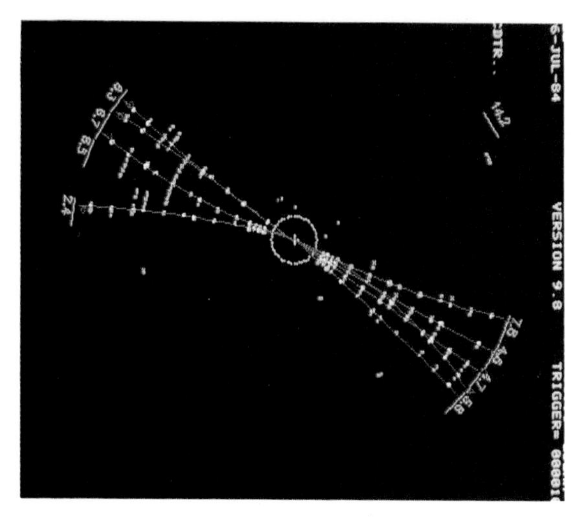

Fig. 10.23 An example of a quark-antiquark back to back 'jet' event seen at the PETRA electron–positron collider in Hamburg, Germany. Most of the tracks are made by pions. This event was observed in the TASSO detector.

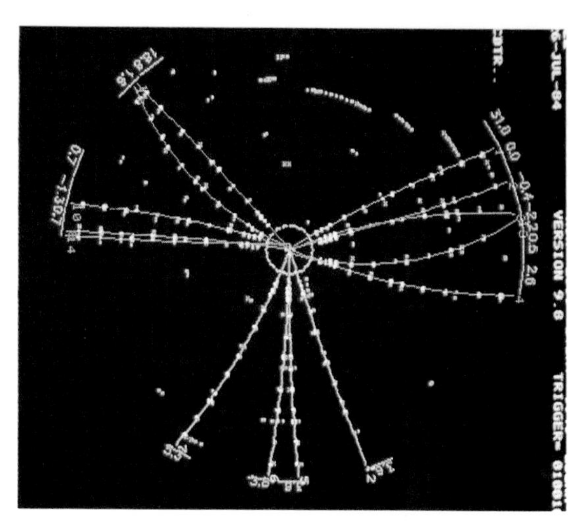

Fig. 10.24 Electron-positron annihilation at PETRA sometimes gives rise to 'three jet' events as in this event seen in the TASSO detector. Such events are believed to be due to the fragmentation into ordinary hadrons of a quark and an antiquark, together with a gluon.

accident and that the interactions between quarks and gluons arrange themselves to make it impossible for us to isolate a single quark. This brings us to the last topic of this book – the problem of 'quark confinement', which is possibly the most important unsolved problem of modern particle physics.

Superconductors, magnetic monopoles and quark confinement

In experiments involving the collision of high energy particles, only ordinary hadrons have ever been observed in the final state. Despite occasional flurries of excitement, no fractionally charged 'quark-like' objects have been conclusively identified. In the collision of two very energetic protons, for example, we do not observe the protons breaking up into quarks. Instead, the collision energy is used to create a whole host of mesons, baryons and antibaryons. Even in a reaction in which we believe an electron is annihilated by a positron to produce a quark and an antiquark going off in opposite directions, we still do not see the quarks. Instead, all that remains is a relic of the initial quark and antiquark motion in the form of two 'jets' of normal hadrons. 'Three-jet events', corresponding to a Feynman diagram in which one of the quarks emits a high energy gluon, have also been seen, but in none of these jets do we see quarks or gluons by themselves.

We now have lots of evidence suggesting that hadrons contain quarks and gluons, yet it seems that their interactions arrange things so that we can never actually isolate an individual quark or gluon. If we try to pull a quark out of a baryon, we have to put in so much energy that we create a quark–antiquark pair. Instead of breaking up the baryon, we end up with a baryon and a meson. We can also now see that Yukawa's meson exchange model of strong interactions is not at all 'fundamental'. Measurements of the contribution of pion exchange to nuclear forces can only tell us very indirectly about the basic quark and gluon force. So how does confinement come about? Nobody knows for sure, but there are some intriguing speculations. One of the most interesting involves a solid-state analogy with superconductors, as for the weak force, but, of course, with a new twist to the argument.

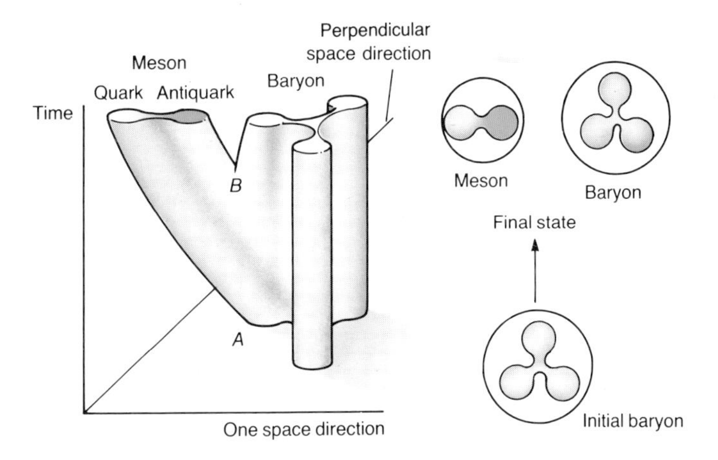

Fig. 10.25 A model for baryons and mesons which illustrates quark confinement. In order to draw the picture, only two space dimensions are shown. A quark is represented as an open circle and an antiquark by a black circle. The gluon forces are shown as an elastic rubber-like sheet that keeps the quarks inside the hadron. As quark A is pulled away from the other quarks, eventually so much energy has been put into the system that a quark–antiquark pair are created at point B and two ordinary hadrons emerge.

There are two new strands of thought that we must introduce. The first concerns the classical theory of electromagnetism. Most people are aware that, although electric charges can exist separately, magnetic charges can only exist as north and south pole pairs, as in a bar magnet. Cutting a magnet in two does not isolate a 'magnetic monopole' but, instead, produces two smaller magnets. This 'magnet analogy' is sometimes used to illustrate a type of 'confinement' – of monopoles in this case – but the mechanism proposed for quark confinement is considerably more subtle. The electric and magnetic fields generated by a system of charges and currents are described by a set of equations known as the Maxwell equations. Because free magnetic monopoles are not found in Nature, these equations are not symmetrical with respect to interchanging electric and magnetic fields. However, if individual magnetic charges and currents did exist, there would be a curious 'dual' symmetry of the resulting equations under the interchange of electric and magnetic fields. It is typical of Dirac's 'quirky' type of originality that he was the first to consider seriously the implications

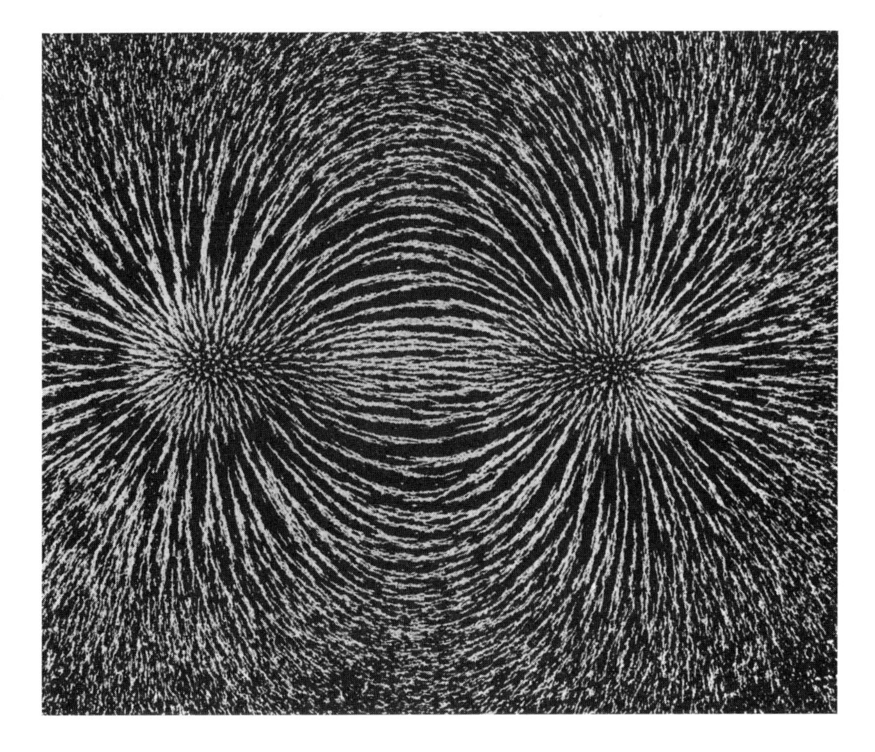

Fig. 10.26 Iron filings scattered on a card above a bar magnet reveal the pattern of the magnetic field around the magnet.

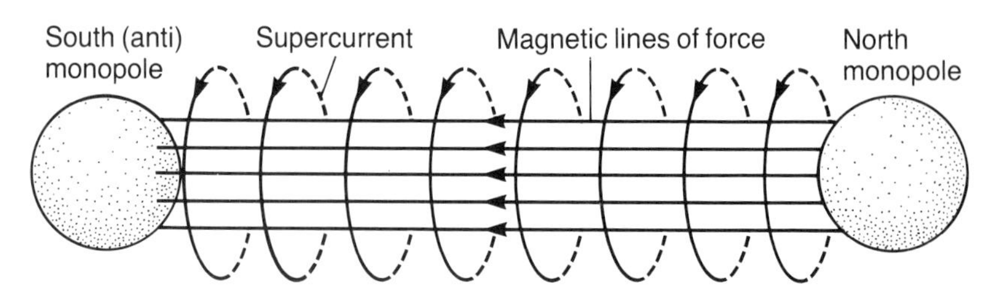

Fig. 10.27 The magnetic field lines joining a magnetic monopole–antimonopole pair in a superconductor. The magnetic field is squeezed into a narrow tube by supercurrents of Cooper pairs circulating around the tube. Quark confinement is thought to be due to a similar mechanism – the ordinary vacuum acts like a 'dual superconductor' in which circulating magnetic monopole currents squeeze the electric field between a quark and an antiquark into a thin tube.

Fig. 10.28 Mr Tompkins inside the proton. In this picture we add an extra chapter to Mr Tompkins' adventures. Inside the proton he meets three mischievous quark dwarfs sentenced to eternal confinement in Gell-Mann's quark prison. Mr Tompkins was appalled at the prison conditions. The dwarfs were all chained together by their ankles but seemed almost oblivious of their cramped confinement. Indeed, they happily maintained they were free, huddled together in the middle of the prison. 'It is positively cosy', they said, but Mr Tompkins was not so sure. 'Now let me show you a trick', said one of the quarks who was standing on his hands. Mr Tompkins and the Professor agreed to hold on to the quark and pull. The quark became more and more excited and kept urging further effort. Suddenly the chain broke and they all tumbled out of prison. When Mr Tompkins had collected his wits he saw an extraordinary sight. Outside the prison were two quarks chained together! Mr Tompkins rubbed his eyes and turned to the Professor who explained that the second quark was actually an antiquark. Mr Tompkins was still bemused and was further surprised to see that back in the prison, there were three happy-looking quarks still chained together. The Professor started to explain that a quark and an antiquark always appear when a gluon chain is broken but Mr Tompkins was no longer listening. It was all too much for him!

of magnetic monopoles for quantum mechanics. By an argument too complicated to attempt here, Dirac showed that the existence of just one quantum mechanical magnetic monopole would imply that all electric charges must be exact multiples of the charge of the electron.

If the above discussion of hypothetical magnetic monopoles seems rather remote from reality, the other new strand in the argument is firmly rooted in experiment. Earlier in this chapter we explained how screening currents act to cancel out any applied magnetic field in a superconductor. Actually, it is found that there are two types of superconductor. 'Type-I' superconductors are those in which the magnetic field is screened out as we have described. In 'type-II' superconductors, on the other hand, the magnetic field is not entirely expelled from the metal, but is allowed to thread its way through in thin 'filaments' (see fig. 8.16). We also encounter another unexpected result of quantum mechanics that is closely related to the Bohm–Aharanov effect described in the opening section of this chapter. This is the fact that the magnetic field threading each filament is 'quantized' and can only have certain values.

We can now explain how quark confinement could come about. Suppose the 'vacuum state' of QCD is like a type-II superconductor. Since QCD is very similar to QED, it should come as no surprise that QCD has both 'colour electric' fields and 'colour magnetic' fields. A vacuum like a type-II superconductor would only allow colour magnetic field to exist in thin tubes as shown in fig. 10.27. Furthermore, the quantized amount of magnetic field allowed through the filament turns out to be just right for the field to begin and end on a colour magnetic monopole. What has this to do with confinement? Fig. 10.26 shows the magnetic field lines of a bar magnet, displayed by the pattern of iron filings. Compared with this pattern, the magnetic field of the monopole–antimonopole pair that threads its way through a type-II superconductor is very different. The electric screening currents circulating around the filament have caused the field lines to be squeezed into a narrow tube. This is just the sort of field pattern that we need for quark confinement. If the field lines are squeezed into a thin tube, then the energy required to separate a quark–antiquark pair will rise in direct proportion to their separation. To separate the quark and antiquark by an infinite distance would therefore require an infinite amount of energy. This amounts to quark confinement.

As promised, there is a final twist to this story. A quark and an antiquark system is bound together by a colour electric field, not by a colour magnetic field. Moreover, quarks do not have a magnetic charge and are not magnetic monopoles. At this point we remind you of the funny electric–magnetic interchange symmetry of Maxwell's equations. Using this 'dual' symmetry, we see that electric fields will be squeezed into the thin tubs required for confinement if the physical vacuum behaves as a 'dual type-II superconductor'. Instead of Cooper pairs encircling filaments containing magnetic fields, the physical vacuum now has magnetic monopole currents trapping tubes of electric fields.

We must conclude by stressing that this mechanism for quark confinement is far from being confirmed. However, it does provide us with a convenient place to end this book. In the earlier chapters, we have seen how the fundamental ideas of de Broglie, Schrödinger, Heisenberg, and all the other great pioneers of quantum physics, can be successfully applied in such diverse fields as stars and superconductors. In this chapter, we have seen how these same ideas are still proving useful in the esoteric discipline of quantum chromodynamics.

Appendix 1

The size of things

Powers of ten

The distance scales that we encounter in the quantum world are much smaller than anything we meet in everyday life. Similarly, the distances involved in any discussions of stars and galaxies are enormously greater than distances between distant places on Earth. A convenient way of writing distances that are either very small or very large is to use a notation which is based on powers of ten.

Large numbers can be written in terms of a power of ten as follows.

$$\text{ten} = 10 = 10^1$$
$$\text{one hundred} = 100 = 10 \times 10 = 10^2$$
$$\text{one thousand} = 1000 = 10 \times 10 \times 10 = 10^3$$

To see how this works, consider how we would write a very large number like the velocity of light in this notation. The velocity of light is usually denoted by the symbol c and is about 300 million metres per second:

$$c = 300\ 000\ 000\ \text{m/s} = 3 \times 10^8\ \text{m/s}$$

The advantage of using the powers of ten notation is evident.

Small numbers can be expressed in a very similar way:

$$\text{one-tenth} = 1/10 = 10^{-1}$$
$$\text{one-hundredth} = 1/100 = 1/10 \times 1/10 = 10^{-2}$$
$$\text{one-thousandth} = 1/1000 = 1/10 \times 1/10 \times 1/10 = 10^{-3}$$

The scale of quantum objects is typified by Planck's constant. This is usually denoted by the symbol h and has a magnitude of about 4.2 thousand trillionths of an 'electron volt second':

$$h = 4.2/1\ 000\ 000\ 000\ 000\ 000\ \text{eV s}$$

In powers of ten notation this is much less cumbersome:

$$h = 4.2 \times 10^{-15}\ \text{eV s}$$

We shall now use this notation to give you some idea of the scale of the various objects that appear in this book.

Mass scales

A familiar unit of mass is the kilogram or 'kg'. This is roughly two pounds in the old non-metric units. Atoms and nuclei are very much lighter than this and a kilogram is a very large and inconvenient unit for such masses. The basic building blocks of atoms and nuclei are protons, neutrons and electrons and these have approximately the masses shown below:

$$\text{proton and neutron mass} = 1.7 \times 10^{-27}\ \text{kg}$$
$$\text{electron mass} = 9.1 \times 10^{-31}\ \text{kg}$$

These very tiny masses contrast with the enormous masses of the planets and stars. Again, a kilogram is not a very convenient unit for such masses as can be seen from the following examples:

$$\text{mass of Earth} = 6 \times 10^{24}\ \text{kg}$$
$$\text{mass of Jupiter} = 2 \times 10^{27}\ \text{kg}$$
$$\text{mass of the Sun} = 2 \times 10^{30}\ \text{kg}$$

Distance scales

In everyday life, the normal unit of length is the metre. This is about the height of a small child and is just over three feet in non-metric units. In Table A1.1 below we contrast the scales of things encountered in this book.

Table A1.1 *Typical distance scales*

m stands for metres

Large scales		**Small scales**	
10^4 m = 10 km	Neutron star, black hole	10^{-4} m	Smallest object visible to naked eye
10^7 m = 10^4 km	White dwarf, Earth	10^{-6} m	Smallest object visible under light microscope
10^8 m = 10^5 km	Jupiter	10^{-8} m	Large molecules
10^9 m = 10^6 km	Sun, normal stars	10^{-10} m	Atoms
10^{11} m = 10^8 km	Red giant, Earth–Sun distance	10^{-14} m	Nucleus
10^{16} m = 1 light year (ly)	Distance travelled by light in one year, distance to nearest stars		
10^{21} m = 10^5 ly	Size of our Galaxy	10^{-15} m	Proton, neutron
10^{21} m = 10^7 ly	Clusters of galaxies		
10^{26} m = 10^{10} ly	Most distant galaxies, quasars		

Fig. A1.1 Relative sizes of stars.

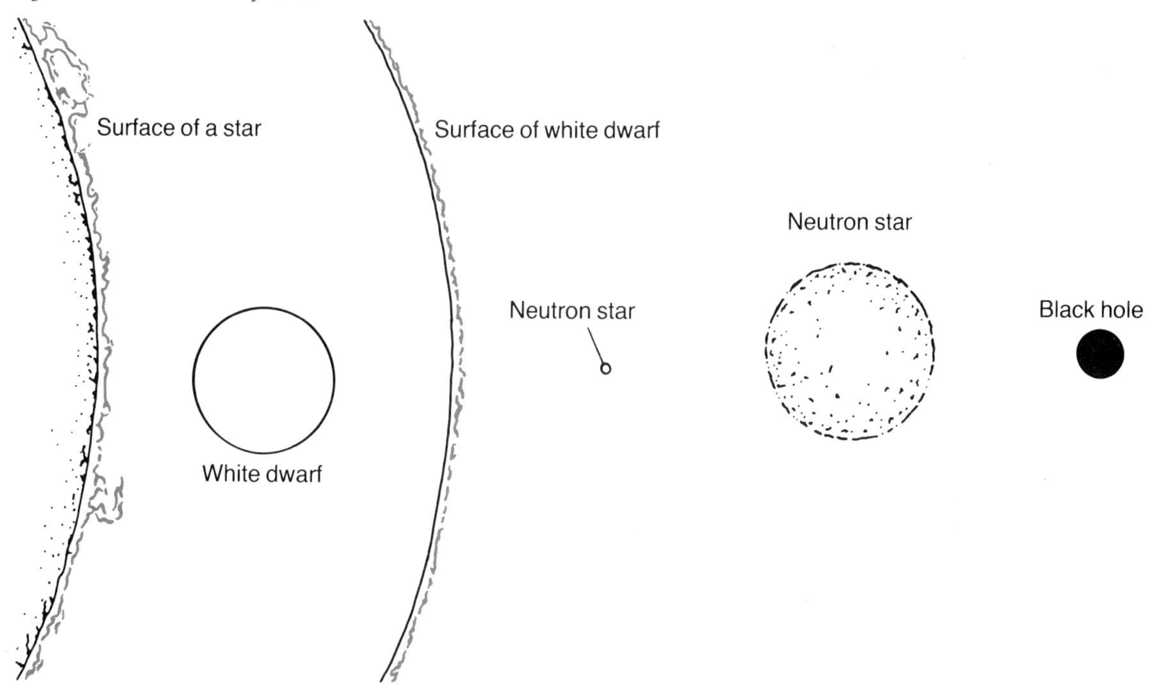

Speed scales

An ordinary walking pace is about one metre per second or about 2.5 miles per hour. Table A1.2 compares the speeds of various objects up to the speed of light, according to Einstein a speed that cannot be exceeded.

Table A1.2 *Speed scales*

m/s stands for metres per second

3 m/s	Reasonable jogging pace
200 m/s	Jet aircraft
300 m/s	Speed of sound in air
10 000 m/s	Escape velocity for rocket
30 000 m/s	Speed of Earth round Sun
2×10^7 m/s	Typical speed of electron in television tube
3×10^8 m/s	Speed of light in vacuum

Electromagnetic spectrum

Light is a type of electromagnetic radiation corresponding to a wave of fluctuating electric and magnetic fields. The wavelengths of visible light form only a very small part of the whole spectrum of electromagnetic radiation. We list in Table A1.3 the typical wavelength ranges of the different types of electromagnetic waves. We have also listed the corresponding photon energy associated with each type of radiation, which reminds us that light interacts in a quantum manner.

Table A1.3 *Electromagnetic spectrum*

m stands for metres; eV stands for electron-volts

Type of radiation	Typical wavelength range	Typical photon energy
Gamma	10^{-11}–10^{-14} m	10^6 eV = 1 MeV
X-rays	10^{-8}–10^{-11} m	10^3 eV = 1 keV
Ultra-violet	4×10^{-7}–10^{-8} m	10 eV
Visible	8×10^{-7} (red)–4×10^{-7} m (violet)	1 eV
Infra-red	10^{-4}–4×10^{-7} m	10^{-1} eV
Microwaves	1–10^{-4} m	10^{-4} eV
Radio	10^{-3} m and longer	10^{-8} eV

Time intervals

A convenient unit for time intervals in normal life is a second – roughly one heartbeat. In Table A1.4 we contrast long and short time intervals.

Table Al.4 *Time intervals*

s stands for seconds

Long times		**Short times**	
10^3 s	Time for light from Sun to reach Earth, neutron lifetime	10^{-6} s	Lifetime of a muon
3.1×10^7 s	One year	10^{-9} s	Typical lifetime of atomic excited state
10^9 s	Human lifetime	10^{-10} s	Time for light to travel one foot (30 cm)
10^{11} s	Origin of city civilizations in the Near East, lifetime of carbon 14	10^{-15} s	Period of oscillation of visible light
10^{14} s	First hominids	10^{-18} s	Time for light to cross atom
10^{17} s	Origin of the solar system, lifetime of uranium 238	10^{-24} s	Time for light to cross nucleus
10^{18} s	Origin of the universe		

Appendix 2

Solving the Schrödinger equation

This appendix is intended for readers with some knowledge of calculus – mainly how to differentiate and integrate sines and cosines. We consider the problem of finding the quantum probability amplitudes for a particle 'in a box'. This involves solving the Schrödinger equation for a particle of some energy E, confined to a certain region by an electrical potential V.

$$E\psi(x) = -\frac{\hbar^2}{2m}\frac{d^2\psi(x)}{dx^2} + V(x)\psi(x)$$

We consider the case where the electron can only move in one space dimension, rather than in three dimensions as in real life. In spite of the apparent artificiality of this situation, this example illustrates many of the features that occur in more realistic problems. A straightforward three-dimensional generalization of this example can be used to illustrate many features of the quantum physics not only of electrons in a metal, but also of neutrons and protons in the nucleus.

The one-dimensional box potential is shown in fig. 4.9. Outside the box the potential is assumed to be infinitely high so that there is no probability of finding the particle in these regions. The particle is thus constrained to be within the box, i.e. the quantum amplitude can only be non-zero between $x = 0$ and $x = L$, the two ends of the box. Inside this region, the particle can move freely: in other words, the quantum amplitude for the particle in the box is determined by the Schrödinger equation with no potential term:

$$E\psi(x) = -\frac{\hbar}{2m}\frac{d^2\psi(x)}{dx^2}$$

To see what form of solution is allowed, we first 'tidy up' the equation by introducing the new variable k:

$$k^2 = 2mE/\hbar^2$$

In terms of k, the Schrödinger equation for the particle in the box now reads

$$\frac{d^2\psi}{dx^2} = -k^2\psi$$

which shows the mathematical structure more clearly. Differentiating the wavefunction twice has to give the same function back again, multiplied by $-k^2$. This equation is well known in classical physics. It is the equation for 'simple harmonic motion' and the solutions are either sines or cosines or a mixture of both of them. Thus, the most general form of the quantum wavefunction is a sum of sine and cosine terms

$$\psi(x) = A\sin(kx) + B\cos(kx)$$

where A and B are arbitrary constants. This is the general form of the solution; what we want is the solution for our particular problem. In our case, therefore, we require the quantum amplitude to vanish for all values of x less than zero and greater than L. We must demand that our solution satisfies these 'boundary conditions'. Imposing the boundary condition at $x=0$

$$\psi(x=0)=A\sin(0)+B\cos(0)=0$$

requires that the constant B must be zero, since the sine of zero vanishes but the cosine of zero does not. The boundary condition at $x=L$ then gives the condition

$$\psi(x=L)=A\sin(kL)=0$$

In order to have a non-zero wavefunction at all, we cannot satisfy this by demanding that A is zero. This therefore implies that $\sin(kL)$ must vanish for an acceptable solution to our problem. This will happen when kL is an integral number times π radians (π radians $=180°$). Thus, the allowed wavefunctions for our problem are those for which k satisfies the condition

$$k=n\pi/L \text{ for } n=1,2,3 \ldots$$

Since k is related to the energy E by the equation

$$E=\hbar^2k^2/2m$$

we see that only certain values of the energy are allowed. We say the energy is 'quantized':

$$E=n^2\frac{\pi^2\hbar^2}{2mL^2} \text{ for } n=1,2,3 \ldots$$

These are the energy levels for the box potential shown in chapter 3 along with their associated sine wavefunctions. All we have done in solving the Schrödinger equation is to determine which wavelengths can fit in a box of length L. This is the origin of energy quantization.

The constant A may be determined as follows. According to Max Born, the probability of finding the particle at any given position within the box is given by the square of the wavefunction at that point. Since the sum of the probabilities of finding the particle at all possible positions within the box must add up to unity, we must have a 'normalization' condition of the form

$$\int_0^L \psi(x)^2 \, dx=1$$

Inserting the wavefunction

$$\psi(x)=A\sin(kx)$$

and using the formula

$$\cos(2kx)=1-2\sin^2(kx)$$

with $k=n\pi/L$, we can perform the integration to obtain the result

$$A^2L/2=1$$

Thus, the constant A is determined to be

$$A=\sqrt{\left(\frac{2}{L}\right)}$$

and the quantum wavefunctions for a particle confined to a box of length L have the form

$$\psi_n(x) = \sqrt{\left(\frac{2}{L}\right)} \quad \sin\,(n\pi x/L) \text{ for } n = 1,2,3\ \ldots$$

The problem of a particle in a three-dimensional box just has wavefunctions with extra sine factors for the y and z directions. In general, however, the mathematics involved in the solution of the Schrödinger equation for more realistic potentials is much more complicated than in this simple example. Nonetheless, this example does contain much of the essential physics.

EPILOGUE

A poet once said 'The whole universe is in a glass of wine'. We will probably never know in what sense he meant that, for poets do not write to be understood. But it is true that if we look at a glass closely enough we see the entire universe. There are the things of physics: the twisting liquid which evaporates depending on the wind and weather, the reflections in the glass, and our imagination adds the atoms. The glass is a distillation of the Earth's rocks, and in its composition we see the secret of the universe's age, and the evolution of the stars. What strange array of chemicals are there in the wine? How did they come to be? There are the ferments, the enzymes, the substrates, and the products. There in wine is found the great generalization: all life is fermentation. Nobody can discover the chemistry of wine without discovering, as did Louis Pasteur, the cause of much disease. How vivid is the claret, pressing its existence into the consciousness that watches it! If our small minds, for some convenience, divide this glass of wine, this universe, into parts – physics, biology, geology, astronomy, psychology, and so on – remember that Nature does not know it! So let us put it all back together, not forgetting ultimately what it is for. Let it give us one more final pleasure: drink it and forget it all!

Richard Feynman

GLOSSARY

alpha particle A helium nucleus which consists of two protons and two neutrons. In alpha radioactive decay an unstable nucleus emits a rapidly moving alpha particle.

amplitude The maximum displacement of a wave motion above its average value. The amplitude determines the amount of energy carried by the wave. Large amplitude waves carry more energy than low amplitude waves as is immediately apparent comparing a stormy sea with calm water. (See *probability* for probability amplitude.)

angular momentum A measure of the amount of rotational motion in a system. Angular momentum is conserved for isolated systems. In quantum theory angular momentum is quantized.

antiparticle A particle with the same mass as the particle concerned but with the opposite 'charge-like' properties. When a particle and its antiparticle meet they can annihilate to energy. Thus, for example, there is an antineutron despite the neutron having zero electric charge, because neutrons have a property corresponding to the conservation of the total number of baryons in the universe. Antineutrons have the opposite value of this 'baryonic charge'.

atom A nucleus with a bound system of electrons. Normal atoms are neutral and form the smallest identifiable amount of a chemical element.

baryon A strongly interacting fermion such as a neutron or proton. Baryons are made up of three quarks.

beta decay A radioactive process in which a neutron (or proton) is changed to a proton (or neutron) emitting an electron (or antielectron). The emitted electron is called a beta particle. Beta radioactive decay proceeds by the weak interaction and is always accompanied by antineutrino (neutrino) emission.

Big Bang model Proposes that the universe began at some definite time (about 20 billion years ago). The early universe is both extremely hot and dense, but is continually cooled by a universal cosmic expansion. This expansion is now observed as the recession of the galaxies.

black hole An object where gravity dominates all other forces and produces a collapse to a singularity where the known laws of physics break down. Once light or anything else enters a critical region about the singularity, at a distance called the Schwarzschild radius, it can never escape. This property gives rise to the name 'black' hole.

boson Particles with the property that any number of them can occupy the same quantum state. The force carrying particles such as photons and gluons are bosons. Also composite objects with an even number of 'matter-like' fermions such as ^{4}He (two protons, two neutrons and two electrons) are also bosons.

charmed quark A quark type, which like the 'up' quark has a charge equal to 2/3 that of proton, but with an additional property 'charm', which is conserved in strong interactions but may be destroyed by the weak force.

'colour' charge Colour charge is a property possessed by quarks and gluons. The colour charge gives rise to the force between quarks (and gluons) in a theory called

quantum chromodynamics (QCD). In ordinary electromagnetic theory the ordinary electric charge plays a similar role to the colour charge in QCD.

conservation When the total amount of some quantity always remains the same it is said to be conserved. For example, the total energy of an isolated system is conserved.

cosmic rays High energy particles originating from extra-terrestrial sources. Some of the relatively low energy cosmic rays undoubtedly come from the Sun, but the origin of the very high energy cosmic rays remains a matter of some controversy.

deuterium An isotope of hydrogen with a nucleus containing one proton and one neutron.

down quark The lightest quark with charge equal to $-1/3$ that of the proton.

Doppler effect The change in the wavelength of waves arriving at a receiver when the source and receiver are in relative motion. The Doppler shift causes an increase in the wavelength when the source and receiver are moving apart and conversely a decrease in wavelength when they approach.

electromagnetic radiation See appendix 1.

electromagnetism One of the fundamental forces of nature which occurs when charged particles interact. The electrical attraction between electrons and protons which holds the atoms together is an example of the electromagnetic force.

electron A negatively charged elementary particle with no strong interaction (with the nucleus). Electrons are constituents of all atoms, surrounding the central nucleus and giving the atom its size, strength and chemical properties. Electrons are very light compared with the nucleus.

electron-volt (eV) A unit of energy comparable with the binding energy of electrons in atoms. An electron-volt is exactly the energy needed to remove an electron from a one volt potential well.

elementary particle A particle without any internal structure and considered to be one of the fundamental building blocks of matter. At the present time quarks and leptons (electrons, neutrinos, etc.) are thought to be elementary.

energy The capability of doing things (technically work). The total amount of energy is conserved though it may be transformed into different forms such as electrical potential or kinetic energy.

exclusion principle (Pauli) No two fermions can have the same quantum numbers. Applied to electrons in atoms it is able to explain the periodic table of the elements.

fermion The elementary 'matter'-like particles or any composite object containing an odd number of elementary fermions. Thus, electrons and quarks are fermions, and so are protons and neutrons because they contain an odd number (3) of quarks. All elementary fermions have spin 1/2.

Feynman diagram A pictorial representation of a contribution to the quantum amplitude for a process built up in terms of simple pieces. Feynman has given definite rules for calculating the amplitude from the simple components of the diagram.

field Any quantity that extends smoothly over an extended region of space. A field should be contrasted with a particle, which we think of as having a well-defined location, rather than being spread out over space.

fission The division of a large nucleus into two roughly equal moderate sized nuclei with possibly some small fragments in addition. Nuclear fission can be induced in certain nuclei by firing neutrons into them and this forms the basis of nuclear reactors.

flavour A property which distinguishes the various types of quark – up, down, strange, charmed, bottom and top. This property is also applied to leptons, where it distinguishes the various types (electron, muon, etc.).

force Anything which causes a change in the motion of a body. There are four fundamental forces: gravity, weak, electromagnetic and strong.

frequency (f) of a wave The number of wave crests passing a particular point in one second. In quantum theory the energy of a quantum particle is proportional to the frequency of the quantum wave.

fusion Any nuclear reaction in which two or more nuclei come together to form a nucleus with more neutrons plus protons than any of the ingredients. Fusion reactions of hydrogen to make helium are the energy source of the Sun.

galaxy A large collection of more than 10 million stars in a region of space well separated from other such collections. Our Sun is one of about 200 billion stars making up our Milky Way Galaxy.

gamma ray A photon of very high energy (see table A1.3 of appendix 1).

gauge theory Electromagnetism is the simplest gauge theory. In the gauge formulation the electromagnetic field arises from the demand that the phase difference of a charged particle quantum wave between different points of space time is unmeasurable. A similar approach can be given for theories of the weak interaction and also the colour force between quarks (QCD).

gluon The quantum particle associated with the 'colour' force between quarks.

ground state The ground state of a quantum system (e.g. atom) is its condition (described by a wavefunction) when it is in its lowest energy level. The quantum condition when the system is in any other energy level is called an excited state.

hadron Any strongly interacting particle. Hadrons are divided into mesons and baryons. Thus, a pion is a meson and a proton is a baryon and both are hadrons.

Higgs particle or boson A hypothetical particle predicted by the Glashow, Salam and Wienberg electroweak theory and needed to give mass to the W and Z particles.

hydrogen The lightest chemical element. The nucleus of ordinary hydrogen consists of a single proton.

ion An atom in which the number of electrons is not equal to the number of protons. An ordinary atom is neutral because the positive charge of the protons in the nucleus is exactly matched by an equal number of oppositely charged electrons surrounding the nucleus.

interference A characteristic property of waves whereby the total wave height of two overlapping waves is the sum of the individual heights. For example, if two identical waves overlap with the crests of one coinciding with the troughs of the other then the two waves will cancel and this is an example of destructive interference. In other situations interference can increase the wave motion.

insulator Material that is a poor conductor of electricity.

isotope Nuclei with the same number of protons but different numbers of neutrons are the isotopes of a given element.

kaon (K-meson) A type of strange meson.

kinetic energy Energy due to the motion of an object.

lambda (Λ) A type of strange baryon.

laser A device for producing coherent light by stimulated emission. The individual 'photon waves' making up coherent light all vibrate and move in step, so we have effectively a single electromagnetic wave.

lepton A fermion such as an electron or neutrino which is not influenced by the strong interaction.

light year The distance travelled by light in one year. In more familiar units it is roughly 10 million million kilometres. Stars in our region of the Galaxy are typically separated by a few light years.

magnetic moment A magnet can be viewed as a pair of magnetic north and south poles separated by a small distance. The magnetic moment is a quantity which describes how the magnet will be influenced by a magnetic field.

magnetic monopole A hypothetical particle consisting of a single isolated magnetic pole.

meson Any strong interacting boson. All mesons are unstable and consist of a bound state of a quark and antiquark.

metal Material which is both a good conductor of heat and electricity and is also shiny (a good reflector of light). In metals, large numbers of electrons are free to move about the whole volume of metal.

molecule A bound system of two or more atoms.

momentum A measure of the quantity of motion defined as the product of mass and velocity. In the absence of external forces the total momentum of the system under consideration is conserved.

muon (μ) A lepton similar to the electron but over 200 times as massive. A muon is unstable and decays to an electron and pair of neutrinos.

neutral current A weak interaction that leaves the charge of the interacting particles unaltered.

neutrino Electrically neutral lepton. Neutrinos interact only by the weak interaction (and gravity) and so (at low energies) are enormously penetrating. Neutrinos come in three varieties called the electron neutrino, muon neutrino and tau neutrino.

neutron Electrically neutral baryon with almost the same mass as the proton. Neutrons together with protons are the constituents of the nucleus.

neutron star A star composed predominantly of neutrons and with a diameter of about ten miles and mass comparable with the Sun. Neutron stars are believed to be formed as a consequence of supernova explosions.

nucleus Dense core of an atom made of neutrons and protons held together by the strong force.

nuclear reaction A collision of nuclei which results in the redistribution of the protons and neutrons so that different nuclei emerge from the collision.

omega minus (Ω^-) The lowest mass baryon made of three strange quarks.

particle A small object with a well-defined location at each instant of time like a bullet. A quantum object sometimes behaves like this but on other occasions behaves more like a wave. In this book we also use the word to describe any quantum object with identifiable properties such as a photon or electron.

periodic table Tabulation of the different kinds of atoms (elements) in terms of increasing number of protons in the nucleus. Atoms with similar chemical and physical properties occur regularly and are placed in columns (see fig. 6.1).

phase A wave motion passing a particular point will undergo an up–down motion. At any given time and place the state of the wave's motion is given by the phase. When one wavelength has passed, the motion will have undergone a complete cycle and this suggests we measure phase by an angle, with 360° corresponding to one cycle. Two waves are in phase when the positions of their crests coincide, otherwise the phase difference expresses the difference in their state of motion.

photon The quantum particle associated with light waves or more generally electromagnetic radiation.

Planck's constant (h) The fundamental constant of quantum mechanics.

plasma A mixture of ions and electrons. In the interior of stars and ordinary ions cannot exist and the plasma is made of nuclei and electrons.

pion (pi-meson) The lightest meson. It comes in three varieties distinguished by an electric charge $+1$, 0 and -1 times that of the proton.

positron Positively charged antiparticle of the electron.

potential energy Energy a system has by virtue of its position or state. For example, the height of an object above the Earth (or other massive object) determines its gravitational potential energy.

probability A number expressing the likelihood of an event taking place. Quantum theory involves intrinsic uncertainty and so probabilities are essential to its description. All possible quantum outcomes have a number associated with them called the probability amplitude. The square of this number gives the probability of the event.

proton A particle with opposite charge to the electron but about 1836 times more massive. Protons experience the strong force and are joined together with neutrons in the nuclei of atoms. The nucleus of ordinary hydrogen consists of one proton.

proton–proton cycle A sequence of nuclear reactions whose overall effect is to convert hydrogen into helium. This sequence is responsible for the generation of energy in the Sun.

pulsar Rapidly varying periodic source of radio waves presumed to be due to a rotating neutron star.

QCD (quantum chromodynamics) The quantum theory of the interaction of quarks and gluons. The quarks and gluons carry a property similar to the electric charge called 'colour' (although it has nothing to do with the familiar concept of colour).

QED (quantum electrodynamics) The quantum theory of electromagnetic interactions.

quantized A physical quantity which can have only certain discrete values in a particular system is described as quantized. Thus, a hydrogen atom has certain discrete energy levels and so energy is quantized (in this system).

quantum number A whole or half integer number or set of numbers which specify the state of a quantum system. For example, the quantized energy levels of a hydrogen atom are denoted by a sequence of positive integers starting with $n = 1$ for the ground state.

quark The elementary particles which are believed to form the basic building blocks of hadronic matter such as protons and pions. Quarks carry a fractional electric charge.

quasar (quasistellar object) A star-like object with a large red shift. If the red shift is interpreted as due to the Hubble expansion of the universe then quasars radiate about 100 times the energy of a conventional galaxy from a central region not much larger than the solar system! Large black holes are involved in most explanations of quasars.

radioactivity Spontaneous disintegration of certain nuclei with the emission of alpha, beta or gamma radiation.

rectifier An electronic component which allows the flow of current in one direction but prevents its passage in the reverse direction.

red shift The displacement of the wavelength of received light to the red end of the spectrum, compared with its emission wavelength. The most common cause of red shift is the Doppler effect due to the source of light moving away from the receiver.

refraction The change in the direction of light or other electromagnetic radiation as it passes from one transparent medium to another.

Schrödinger equation The basic equation of quantum mechanics which describes the behaviour of a particle in a potential.

semiconductor A material whose ability to conduct electric current is intermediate between metals and insulators and also increases with increasing temperature.

spin Fundamental property of a particle corresponding to an intrinsic angular momentum.

spontaneous emission Photons emitted from an isolated atom (or other quantum system) in a transition from an exicted state to a lower energy state.

strangeness Property of hadrons conserved in the strong interaction but destroyed by the weak interaction. It is associated with the strange quark.

strong interaction (force) The force that holds the nucleus together and is responsible for the interaction of hadrons. The strong force is a remnant of the inter-quark 'colour' force.

superconductor A material (usually a metal or alloy) whose electrical resistance vanishes below a certain critical transition temperature.

superfluid A fluid that flows without friction and has a high thermal conductivity below a certain critical temperature.

supernova A catastrophic stellar explosion in which the power output becomes comparable with that of a whole galaxy for about a month and thereafter gradually declines. The explosion may leave behind a neutron star or black hole.

temperature A measure of hotness. If two bodies in contact have different temperatures then heat will flow from the hotter body to the colder one. The temperature of a body is a measure of the average kinetic energy of its atoms (or other constituents).

transistor A semiconductor device in which the flow of current between two contacts can be controlled by a potential applied to a third contact.

tunnelling The ability of quantum objects to pass through regions which classically are energetically forbidden.

uncertainty principle (Heisenberg) If an experiment specifies the position of a particle to a certain accuracy (Δx) then it automatically results in the momentum becoming unpredictable to an accuracy greater than Δp, such that the product $\Delta x \Delta p$ is greater than a certain minimum value of the order of Planck's constant. A similar principle applies to the uncertainties in energy and time measurements.

up quark The lightest quark with charge equal to 2/3 that of the proton.

virtual particle A particle violating the conservation of energy and existing only for a brief instant of time so the energy–time uncertainty principle is satisfied.

W particle Massive charged particle that together with the neutral Z particle is associated with the weak interaction just as the photon is associated with the electromagnetic interaction.

wave motion Any kind of oscillating or moving disturbance which at any given instant is spread over a region of space. The simplest kind of wave corresponds to a periodic up–down disturbance. The distance between adjacent crests is called the wavelength.

weak force One of the fundamental interactions. The weak force is responsible for beta decay and any interaction involving neutrinos.

white dwarf A dense compact remnant star with a typical mass comparable with that of the Sun but of a size comparable with that of the Earth. The star is held up by the Pauli principle applied to electrons and is still hot but is gradually cooling.

X-rays Relatively energetic form of electromagnetic radiation or photons (see appendix 1).

Z particle Neutral massive boson associated with the weak interaction.

zero-point motion Vibrational motion of atoms at absolute zero temperature due to Heisenberg's uncertainty principle.

Zeeman effect Splitting of single spectral lines into two or more components when the atoms are subject to a magnetic field.

SUGGESTIONS FOR FURTHER READING

Quantum mechanics

R. P. Feynman (1965). *The Character of Physical Law* (MIT Press)

This book comprises seven lectures presented at Cornell University in 1964. Even 20 years on, the book still sparkles with Feynman's unique style and wit.

R. P. Feynman (1965). *The Feynman Lectures on Physics* (Addison-Wesley)

All three volumes are notable for their insight and novel presentation of all aspects of physics. Volume 3 contains Feynman's unusual approach to quantum mechanics – most students find it rather difficult and prefer a more conventional approach.

R. P. Feynman (1985). *QED* (Princeton University Press)

An entertaining yet serious attempt at a 'popular' exposition of quantum electrodynamics or QED. As ever, Feynman tries to achieve maximum clarity and simplicity without compromise by distortion of the truth.

A. P. French and E. F. Taylor (1978). *An Introduction to Quantum Physics* (Norton, USA: Nelson, UK)

A traditional quantum mechanics book, but more wordy than most, and quite accessible for the majority of its chapters.

J. C. Polkinghorne (1984). *The Quantum World* (Longman)

A clear and careful introduction to the conceptual problems of quantum mechanics. The famous paradoxes of Schrödinger's cat, Wigner's friend and of Einstein, Podolsky and Rosen are all discussed in detail.

F. E. Close (1983). *The Cosmic Onion* (Heinemann)

The most up-to-date and readable account of the recent advances in our understanding of the forces of Nature and elementary particles.

G. Gamow (1965). *Mr Tompkins in Paperback* (Cambridge University Press)

The famous physicist George Gamow's entertaining account of Mr Tompkins' imagined explorations of relativity and quantum mechanics.

Historical background

O. Frisch (1979). *What Little I Remember* (Cambridge University Press)

E. Segré (1980). *From X-rays to Quarks* (Freeman)

These two fascinating autobiographical accounts of the early days of quantum mechanics are well worth reading.

R. P. Feynman (1985). *Surely You're Joking, Mr Feynman!* (Norton)

Feynman's delightful and entertaining autobiography containing many of the legendary 'Feynman stories' and much else besides.

A. Pais (1982). *Subtle is the Lord – The Science and the Life of Albert Einstein* (Oxford University Press)

Probably the definitive book on Einstein's contributions to the foundations of quantum mechanics and his development of the general theory of relativity.

P. Goodchild (1980). *J. Oppenheimer – Shatterer of Worlds* (BBC Publications)

The book from the BBC television series about a fascinating piece of contemporary history.

S. Augarten *Bit by Bit – An Illustrated History of Computers* (Tickner and Fields)

An absorbing account of the history of computing from the early pioneers like John Von Neumann and Alan Turing, to modern heroes like Jobs and Wozniak, the creators of the personal computer.

PHOTO-CREDITS

We would like to express our thanks to the following individuals and organizations for allowing us to use their material in this book:

Prof. J. F. Allen: figs 8.13, 8.14; American Institute of Physics, Niels Bohr Library: p.5 Born, p.106 Chandrasekar; Prof. C. D. Anderson: fig. 9.3; Ann Ronan Picture Library: p.75 Mendeleev; Argonne National Laboratory: p.76 Fermi; Associated Press: p.76 Pauli, p.88 Shockley, Brattain and Bardeen, p.112 Maiman, p.142 Yukawa; Prof. J. Bardeen: p.120 Bardeen, Cooper and Schrieffer; Cyril Band: fig. 8.12 from Mendelssohn, K. *Cryophysics*. Interscience; A T & T Bell Laboratories; fig. 6.18; G. Binnig and H. Rohrer: fig. 5.8 (earlier published in *Scientific American* 253 (2), p.54 (1985), British Broadcasting Corporation: fig. 5.25, p.116 Einstein (BBC Hulton Picture Library); Brookhaven National Laboratory: figs 7.3, 10.7, 10.8, p.153 Ting; Roger Cashmore and the TASSO collaboration: figs 10.23, 10.24; Cavendish Laboratory, University of Cambridge: p.5 Thomson, p.16 Heisenberg, p.37 Rutherford, figs 5.11, 7.11 (Mullard Radio Observatory), p.121 Josephson, p.137 Maxwell; CERN: figs 9.4, 10.4, 10.6, 10.9, 10.10, 10.12–10.16, 10.19, 10.20; Chicago Historical Society: fig. 5.17; Cornell University Department of Physics: p.99 Bethe; A. V. Crewe *et al. Science* 168, pp. 1338–40 © 1970: fig. 3.5; The *Daily Mail*: fig. 5.18; Copyright Earth Satellite Corporation under the GEOPIC trademark: fig. 6.11; Education Development Center. Inc., Newton MA: figs 1.1, 4.7, 5.5; Prof. R. Feynman: p.21 Feynman; Fiat Auto: fig. 8.4; Prof. T. H. Geballe: fig. 8.18; Prof. M. Gell-Mann: p.34 Gell-Mann; Harvard University Department of Physics: p.147 Glashow and Weinberg; Prof. S. Hawking: p.136 Hawking; Dr Robert Herman and Dr R. A. Alpher: p.132 Gamow; Hilbert, Constance Reid, Springer-Verlag © 1970: p.140 Weyl; Hiroshima–Nagasaki Publishing Committee (Eiichi Matsumoto): fig. 5.19; Prof. Sir Fred Hoyle: p.101 Hoyle; Dr C. M. Hutchins: fig. 4.8; IBM Corporation: figs 6.24, 8.17; INMOS Ltd: figs 6.23, 6.25; Institute for Advanced Study, Princeton NJ: p.92 Oppenheimer and von Neuman; The Institute of Physics: fig. 10.3 from *Rep. Prog. Phys.* 13, 350 (1950); Intel: fig. 6.20; John Wiley & Sons Inc: fig. 8.9 from Smith *Principles of Holography* © 1969; Copyright Prof. Dr. C. Jönsson: fig. 1.10; Kalmbach Publishing Co., Milwaukee WI: fig. 7.5 from *Astronomy Magazine* with permission (artwork by John Clarke); Karl Kuhn and J. S. Faughn: fig. 10.26 from *Physics in Your World*; Lawrence Berkeley Laboratory, University of California: p.62 Livingstone and Lawrence, fig. 9.5; E. Leitz Inc: fig. 6.12; Lick Observatory: figs 7.7, 7.8, 7.12; Macdonald & Co. (Publishers) Ltd: fig. 9.15; Ministry of Agriculture, Fisheries and Food (SEM Unit, Slough Laboratories): fig. 3.7 crown copyright; Rajat Mitra: p.115 Bose; Mount Wilson and Las Campanas Observatories, Carnegie Institution of Washington: figs 2.7, 7.6, 7.9, 7.10 (nebula picture), 9.12; Dr E. W. Mueller: fig. 5.7; Mullards, Southampton (Roger Pearce): fig. 6.21; Museum Boerhaave Leiden: p.188 Onnes; NASA: figs. 1.2, 4.5, 7.2 (montage by Stephen Meszaros), 8.1; National Optical Astronomy Observatories: figs 2.5, 9.11; National Portrait Gallery: p.27 de Broglie, p.97 Eddington, p.115 Gabor; Naval Research Laboratory, Washington DC: fig. 4.5; The New York Academy of Sciences: fig. 1.12 from Dr Hannes Lichte (the experiment was conducted using a Mollenstedt-type electron biprism interferometer); The Slide Centre: fig. 5.22; Niels Bohr Archive, Copenhagen: p.124 Dirac and Heisenberg; Niels Bohr Institute: p.39 Bohr; Copyright The Open University Press: fig. 1.6 from *Discovering Physics* (S271), fig. 4.6; © Oxford University Press, 1961: figs 1.4, 1.8, 2.4, 3.1; Pfaundler, Innsbruck: p.31 Schrödinger; Photo Deutsches Museum Munchen: p.17 Planck, p.136 Fraunhofer; Physicians for Social Responsibility: fig. 5.20; Proceedings of The Royal Society: fig. 5.10; Prof. G. D. Rochester, FRS and Sir Clifford Butler, FRS: fig. 10.5; Albert Rose: fig. 2.6 from monograph *Vision: Human and Electronic*; Prof. Thomas D. Rossing: fig. 4.10; Royal Observatory Edinburgh: fig. 4.14 © 1982, p.107 Bell Burnell; Rutherford Appleton Laboratory: fig. 5.12; Prof. Abdus Salam: p.148 Salam, fig. 10.11; Schoken

Books Inc: figs. 3.10, 6.10 from Scharf, D. *Magnifications* © 1977: Science Photo Library: fig. 3.6 from T. Brain; *Scientific American*: fig. 5.23; Prof. Emilio Segre: p.127 Anderson and Glaser; Maurice M. Shapiro, Colorado Associated Press: p.60 Gamow; SLAC: figs 3.8, 3.9, 3.20, p.155 Goldhaber, Perl and Richter; Roger Stalley: p.31 wall plaque; Texas Instruments: fig. 6.19; Prof. Gerard t'Hooft: p.157 t'Hooft; Times Mirror Magazines Inc: fig. 6.22 reprinted from *popular Science* with permission © 1946; Charles H. Townes: p.109 Townes; Trustees of the British Museum: p.1 Newton, p.3 Young, fig. 7.10 (Chiense script); Ullstein Bilderdienst: p.61 Cockroft, Walton and Rutherford, p.67 Meitner and Hahn; US Atomic Energy Commission: figs 4.4, 5.21; Richard F. Voss: fig. 2.11; Prof. C.N. Yang: p.141 Yang; Prof. G. Zweig: p.34 Zweig.

We have been unable to trace the copyright holders of the following figures: figs 1.3, 2.2, 5.3, 5.4, 6.1, p.87 Stern, p.91 Hartree, p.123 Einstein.

Sources for Feynman quotes

Prologue: *Feynman Lectures in Physics* vol. 1, chap. 3, p.6; Epilogue: *Feynman Lectures in Physics* vol. 1, chap. 3, p.10; pp.1, 13: Feynman, R. (1967) *Character of Physical Law*, chap. 6. MIT Press; p.27: *Feynman Lectures in Physics* vol. 3, chap. 16, p.12; p.36: *Feynman Lectures in Physics* vol. 3, chap. 2, p.6; p.53: *Feynman Lectures in Physics* vol. 3, chap. 8, p.12; p.74: *Feynman Lectures in Physics* vol. 3, chap. 2, p.7; p.96: *Feynman Lectures in Physics* vol. 1, chap. 3, p.7; p.109: *Feynman Lectures in Physics* vol. 3, chap. 21, p.1; p.123: Feynman, R. (1949) *Phys. Rev.* 76, 749; p.137: Feynman, R. from *Horizon* BBC2 produced by C. Sykes. Reprinted in the *Listener*, 26 November, 1981, p.639.

SUBJECT INDEX

NAME INDEX

0 1138 0108048 3
Wentworth - Alumni Library